Biomedical Implants

This book provides a comprehensive overview of the development of implants, from the selection of materials to the outcome of the process. It covers various steps, including biocompatible material, synthesis, and characterization, compatibility and limitations of materials, specific implants, and finite element analysis of medical implants. It also presents a comparison between predictions and experimental results by studying real-world problems and addresses the issue of sustainability in implant manufacturing, process modeling, and optimization in additive manufacturing supported by case studies.

Features:

- Covers the development of implants from the selection of material to the suitable process of manufacturing technologies.
- Includes biocompatible material, synthesis, characterization, compatibility, and limitations of materials.
- Reviews biofabrication in terms of artificial organs and soft tissues.
- Discusses implant manufacturing, including additive and micro-manufacturing and failure analysis through case studies.
- Addresses the issue of sustainability in implant manufacturing.

This book is intended for researchers and graduate students specializing in mechanical, biomedical, healthcare engineering, biomaterials, and additive manufacturing.

Biomedical Implants

Materials, Design, and Manufacturing

Edited by
Ravi K. Dwivedi, Premanand S. Chauhan,
Avadesh K. Sharma, Madhavi Singh,
and Anupma Agarwal

CRC Press
Taylor & Francis Group
Boca Raton London New York

CRC Press is an imprint of the
Taylor & Francis Group, an **Informa** business

Designed cover image: shutterstock

First edition published 2024

by CRC Press
2385 NW Executive Center Drive, Suite 320, Boca Raton FL 33431

and by CRC Press
4 Park Square, Milton Park, Abingdon, Oxon, OX14 4RN

CRC Press is an imprint of Taylor & Francis Group, LLC

ISBN: 9781032428406 (hbk)
ISBN: 9781032450377 (pbk)
ISBN: 9781003375098 (ebk)

DOI: 10.1201/9781003375098

Typeset in Times
by Deanta Global Publishing Services, Chennai, India

This book is dedicated to our mentor

Professor H B Khurasia

*Whose guidance and unwavering support have
been a constant source of inspiration.*

*This incredible journey of knowledge and discovery
is the consequence of your wisdom.*

*Thank you for paving the path of knowledge and
shaping our understanding of the world.*

Your mentorship will forever be cherished in the pages of this work.

Contents

Preface

In this era of advancement, technology is growing exponentially. Every aspect of human life is deeply touched by the abruptly changing technology. Technological advancement is more focused on providing customized solutions to problems, primarily related to the medical field. This advancement is also visible in the development of implants. Currently, implants are being used in different settings as joint replacements, dental implants, and different cardiovascular structures. Most commonly, implants are being used for bones. Bone is a complex tissue with self-remodeling attributes supporting self-healing. However, this natural healing is though and limited to a particular size and if the defect is too large or non-healable at self-pace, an outside intervention in terms of an implant is needed.

The development of patient-specific implants with a high satisfaction rate is the prime interest of researchers and industrialists. This development of patient-specific implants demands biocompatible material and efficient technologies to customize the desired products. The use of titanium, stainless steel, and cobalt-based alloys as implant material has a long history. Metal alloys are not very flexible but their biocompatibility, strength, and high corrosion resistance make them more suitable for use as an implant material. Lack of flexibility leads to stress. The use of new generation materials, such as Ti6Al4V and Co-Cr-Mo that have moduli more than the original bone, with suitable manufacturing techniques facilitate implant technology and help to customize patient-specific implants with zero stress shielding. While designing porous implants, choosing the appropriate material matching the porosity, pore size, and interconnectivity should be of the utmost priority in an attempt to make them compatible with the bone and surrounding tissues. The development of more precise and patient-specific implants is also dependent on the selection of suitable materials and manufacturing techniques.

In the past, various conventional methods were used for manufacturing implants but each of the methods had one or more limitations. To overcome the shortcomings of conventional methods, for the past decade, additive manufacturing (AM) has taken over the patient-specific implant manufacturing market, due to its ability for precise customization and a variety of printing techniques. Additive manufacturing techniques also provide the freedom to manufacture any object with slight changes to printing principles, thereby widening its scope over other conventional manufacturing methods and attracting budding researchers, evidence by the many new variants of AM techniques. The compatibility of various AM techniques with a wide range of materials shows its resilience when choosing material during manufacturing. AM technologies have also brought notable changes to implant manufacturing by reducing the steps used in conventional manufacturing methods and preparing the desired implant model with ultimate precision and extreme feature accuracy which further increases the customer's satisfaction rate. AM technologies are facilitated through computer-aided design (CAD). CAD makes it easy to modify or change the design at any stage, before manufacturing. A detailed analysis and simulation is made easier using other software and ensures zero errors before proceeding with manufacturing.

This book focuses on the development of implants from the selection of material to the final outcome following various steps. The book includes the study of bio-compatible material, synthesis, and characterization, compatibility and limitations of materials, medical imagination for patient-specific implants, and finite element analysis of medical implants. It also includes the implant manufacturing journey from conventional to additive manufacturing technologies, the scope of micro-manufacturing in implant manufacturing processes, and failure analysis through case studies. It also presents a comparison between predictions and experimental results by studying real-world problems, addresses the issue of sustainability in implant manufacturing, process modeling, optimization in additive manufacturing, biofabrication: artificial organs and soft tissues, and the capabilities and limitations of materials and manufacturing processes in medical implants. The last chapter of the book concludes the studies presented.

The book aims to facilitate researchers and industrialists working in the field of implant manufacturing while incorporating a comprehensive study on the development of implants from the selection of material to the suitable process of manufacturing technologies. It presents state-of-the-art research in the thematic area by systemizing the literature from raw material (compatible with implant manufacturing), manufacturing tools, and techniques to the expected outcome of the technologies used through real-world case studies. This book would be a great help to those who want their start-up to be a small implant manufacturing unit.

Editors

Ravi K. Dwivedi is a professor and head of the Centre of Excellence in Product Design and Smart Manufacturing, MANIT, Bhopal. He has over 44 years of experience, including 15 years in the Indian Air Force, 1.2 years in the industry, and 28.2 years in teaching. He has authored 2 books, 11 book chapters, and 2 edited books, and was editor of the proceedings for many international conferences and granted patents. He has guided many scholars of PhD, MTech, and undergraduates. His area of research is additive manufacturing. He has been granted many funded projects and completed several industrial consultancies. He is also an executive member of many professional societies. He is a technical expert member of many government bodies and universities.

Premanand S. Chauhan works as a director at Sushila Devi Bansal College of Technology, Indore. He has 7 years of industry experience and 20 years of teaching experience. He has authored 2 books, 55 research articles, and has published three patents and one patent granted. He is the editor of the proceedings of many reputed international conferences. He is the recipient of the Exemplary Academic Administrators of Higher Education Institutes across India in 2019 conferred by the Academic Council of uLektz and Director of the Year 2022 awarded by the Universal Mentors Association, India. He supervised more than 15 MTech theses and 4 PhD scholars are working under his supervision. His area of research is product and process design. He is also the technical adviser for many industries working in the field of manufacturing.

Avadesh K. Sharma is an associate professor and head at the Department of Mechanical Engineering, Rajkiya Engineering College, Mainpuri. He has 21 years of teaching and 12 years of research experience. He has published more than 75 research papers in reputed international and national journals and conferences, authored 1 book, and edited the proceedings of international conferences. Sharma has guided more than 18 MTech theses and 2 PhD students. He is also a reviewer for various national and international journals. He is a member of the BOS of Dr APJ AKTU Lucknow and IET Lucknow. His area of research is mechanical vibrations and materials.

Madhavi Singh doctor in family medicine and fellow in hospital medicine is currently associate professor in the Department of Family and Community Medicine at Penn State College of Medicine. Her areas of interest are varied encompassing a wide scope of family medicine including new advances in medical care. She is one of the editors of the IC4M2022 conference proceedings.

Singh completed her medical education in Gajra Raja Medical College, Gwalior, India where she also completed her Obstetrics and Gynecology postgraduate degree. She later completed her family medicine residency at the University of Pittsburgh Medical Centre, Altoona.

Anupma Agarwal is an assistant professor in the School of Mechatronics Engineering at Symbiosis University of Applied Sciences, Indore. She has 16 years of experience in teaching. She has published two patents and edited the proceedings of four international conferences. She authored 19 papers and guided three MTech theses. Her area of research is atmospheric physics and materials.

Contributors

Anupma Agarwal
Symbiosis University of Applied
Sciences, India

Neha Agrawal
Banaras Hindu University, India

Ragavi Alagarsamy
VMMC and Safdarjung Hospital, India

Azan Ali
Swinburne University of Technology,
Malaysia

Harakh Chand Baranwal
Banaras Hindu University, India

Prateek Behera
All India Institute of Medical Sciences,
India

Abhinav Bhagat
All India Institute of Medical Sciences,
India

Pranav Charkha
DY Patil University, India

Premanand S. Chauhan
S.D. Bansal College of Technology,
India

Sushil Kumar Choudhary
Central University, India

Boppana V. Chowdary
The University of the West Indies,
St Augustine Campus, Trinidad and
Tobago

Ravi Kumar Dwivedi
Maulana Azad National Institute of
Technology Bhopal, India

Vishal Francis
Lovely Professional University, India

Nishkal George
The University of the West Indies,
St Augustine Campus, Trinidad and
Tobago

Sankalp Gour
Maulana Azad National Institute of
Technology, India

Radha Sarawagi Gupta
All India Institute of Medical Sciences,
India

R.S. Jadoun
G.B. Pant University of Agriculture &
Technology, India

Santosh Jaju
G.H. Raisoni College of Engineering,
India

Elammaran Jayamani
Swinburne University of Technology,
Malaysia

Farheen Khan
Maulana Azad National Institute of
Technology, India

Amit Khemka
Banaras Hindu University,
India

Ashutosh Kumar
Maulana Azad National Institute of
Technology, India

Deepak Kumar
Maulana Azad National Institute of
Technology, India

Sudhanshu Kumar
Maulana Azad National Institute of
Technology, India

Virendar Kumar
Harcourt Butler Technical University,
India

Babu Lal
All India Institute of Medical Sciences,
India

Lakha Mattoo
The University of the West Indies,
St Augustine Campus, Trinidad and
Tobago

Vyanktesh Naidu
G.H. Raisoni College of Engineering,
India

Ankit Nayak
The University of Hong Kong, Hong
Kong

John Ashutosh Santoshi
All India Institute of Medical Sciences,
India

Abhishek Sharma
G.L. Bajaj Institute of Technology and
Management, India

Ajay K. Sharma
Institute of Engineering and
Technology, India

Avadesh K. Sharma
Rajkiya Engineering College, India

S.S.P.M. Sharma B
Symbiosis University of Applied
Sciences, India

Madhavi Singh
Penn State College of Medicine,
Hershey, PA

Nishant K. Singh
Harcourt Butler Technical University,
India

Rajeev Singh
Rajkiya Engineering College, India

Yashvir Singh
Graphic Era Deemed to be University,
India

KokHeng Soon
Swinburne University of Technology,
Malaysia

Ayush Srivastav
Ola Electric Technologies Pvt. Ltd., India

Mohammad Taufik
Maulana Azad National Institute of
Technology, India

Rajeev K. Upadhyay
Hindustan College of Science and
Technology, India

Mohammad Usman
Swinburne University of Technology,
Malaysia

Hridwin Vishaal
Chaitanya Bharathi Institute of
Technology, India

Mohamad Kahar bin Ab Wahab
Universiti Malaysia Perlis, Malaysia

Vinod Yadav
Maulana Azad National Institute of
Technology, India

Rachna Zirath
Upstate Medical University, Syracuse,
NY

1 Introduction to Biocompatible Materials

Neha Agrawal, Amit Khemka,
Harakh Chand Baranwal, and Madhavi Singh

1.1 INTRODUCTION

Biomedical implants composed of natural or derived materials are increasingly being used to replace or improve the functioning of various biological structures.

In the aging population, osteoarthritis is a very common problem leading to significant debilitation [1]. Bioimplants in the form of joint replacements serve as a crucial treatment providing remarkable improvement in quality of life. Sports injuries in young athletes can lead to replacements. This requirement for joint replacement and rehabilitation has become one of the main driving forces for continuous research and development in the field of biomaterials for biomedical implants.

Cardiovascular diseases also require biomaterials, especially in cases of vascular grafts, stents, and pacemakers. Biomedical implants are often used in dental applications, ophthalmology, craniofacial, plastic, and reconstructive surgery, and veterinary medicine (Figure 1.1).

The area of biomaterials is not novel and can be dated back 4000 years [2]. In that era, the Egyptians and the Romans used linen instead of suture material. They also used gold and iron in various dental applications but without any awareness of the complications of corrosion. The pioneering discovery of the origin and evolution of bioimplants was made by researchers such as Harold Ridley, Paul Winchell, Per-Ingvar Branemark, Otto Wichterle, and John Charnley. After World War II, various other materials such as nylon, Teflon, silicone, stainless steel, and titanium came into existence and gradually started to be used. Currently, materials such as metals, polymers, ceramics, and composites are being used on a wide scale to fabricate bioimplants [3].

1.2 DEFINITION

"Biocompatibility of a long-term implantable medical device refers to the ability of the device to perform its intended function with the desired degree of incorporation in the host, without eliciting any undesirable local or systemic effects in the host" [4]. Therefore, acceptability by the human body is the principal requisite for any material to be used as a biomaterial [5]. The implant material placed in the human system

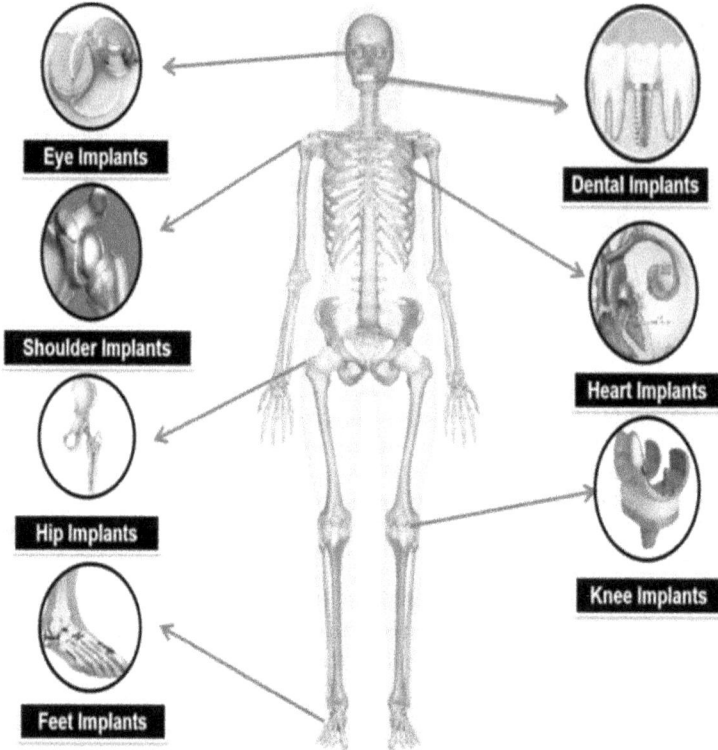

FIGURE 1.1 Various applications of biomedical implants.

should not cause any detrimental effects such as allergic reactions, inflammation, or toxicity, either just after implantation or postoperatively.

The other most desirable property is that the biomaterials should not undergo fracture and that they have adequate mechanical strength to endure the forces to which they are subjected. Furthermore, a biomaterial should possess high corrosion and wear resistance in the presence of an extremely corrosive body environment, in addition to mechanical strength and fracture toughness. Biomaterials should also have properties that allow them to remain viable for a long period of time. This limits the complications associated with their replacement.

1.3 PROPERTIES OF BIOMATERIALS

With the need for viability as per the factors described above, biomaterials for biomedical implants should be carefully chosen considering four major types of properties:

(1) bulk properties,
(2) surface properties,
(3) chemical properties,
(4) tribological properties.

1.3.1 BULK PROPERTIES

The fracture strength, yield strength, elastic modulus, and creep are the principal properties of a biomaterial that need to be carefully evaluated before using the material for load-bearing purposes in hard tissue implants.

1.3.1.1 Modulus of Elasticity

The elastic modulus determines the stiffness or rigidity of a material. Biomaterials with a comparatively similar modulus of elasticity to that of bone must be chosen to maximize the homogeneous dissemination of stress at the implant surface as well as to prevent the corresponding motion at the implant–bone junction.

1.3.1.2 Tensile, Compressive, and Shear Strength

The tensile strength [6] refers to the maximum force a material can resist when stretched before it undergoes permanent deformation. The compressive strength measures material's capability to resist compression. The biomaterial should have high tensile and compressive strength to prevent breakage and improve its steadiness while in use. These properties help keep the resultant stresses in the implant low even when the stress transfer increases.

1.3.1.3 Yield Strength

The yield strength determines the maximum stress at which there is a transition of a material from elastic toplastic. High yield strength implies that the biomaterial can withstand high stresses under cyclic loading without any brittle fracture.

1. **Ductility:** Ductility is a critical property for the mechanical compliance of a material. It is desirable for a biomaterial to have high ductility for contouring and shaping when used as a biomedical implant.
2. **Hardness:** Hardness determines the suitability of a biomaterial for various clinical applications. It is desirable for an implanted biomaterial to possess hardness equal to that of the bone. With the increase in the hardness of a biomaterial, its wear resistance increases, thereby increasing its survival period.

1.3.2 SURFACE PROPERTIES

1.3.2.1 Surface Energy and Surface Tension

The surface tension of a liquid and the surface energy of a joined part measure the amount of wetting of a material by an adhesive. The less surface tension, the more-wettability and improved osteoblastic [7] adhesion.

1.3.2.2 Surface Roughness

Changes in the grain particle size and alterations to the surface roughness of a bio-implant affect the mechanical properties, wear resistance, and corrosion resistance and also influence biocompatibility. Ultrafine grain exhibits stronger mechanical properties compared to coarse-grained biomaterials. This is as a result of ultrafine-grained particles having more surface atoms, thereby increasing the surface area of the implant and enhancing the attachment of the cell to bone. The surfaces of implants are categorized on the basis of various characteristics such as surface coarseness or roughness, texture, and the orientation of irregularities, as discussed below.

1. According to the surface roughness, Wennerberg and coworkers classified implant surfaces as
   a. Minimally rough (0.5–1 m).
   b. Intermediately rough (1–2 m).
   c. Rough (2–3 m).
2. According to their texture, the implant surface can be further categorized as
   a. A concave texture that is mainly produced through additive treatments such as titanium plasma spraying and hydroxyapatite (HA) coating [6].
   b. A convex texture that is mainly fabricated through a subtractive treatment such as etching and blasting.
3. According to the orientation of surface irregularities, implant surfaces can be classified as
   a. Isotropic surfaces.
   b. Anisotropic surfaces.

1.3.3 CHEMICAL PROPERTIES

Corrosion is defined as the deterioration of a metal resulting from a reaction with its environment. It affects the duration of the service period as well as the life span of implants composed of metals and alloys [8]. Resistance against corrosion is a critical requirement for the successful implementation of biomaterials, especially when used as biomedical implants in cases of bone fractures and replacements. The following is a list of different types of corrosion.

1.3.3.1 Crevice Corrosion

Crevice corrosion is a form of local corrosion that mainly occurs in narrow regions. It occurs in regions of contact such as a metal–metal interface or a metal–non-metal interface. In the case of implants, it mainly occurs at the implant screw–bone interface. In crevice corrosion, once the passive layer is destroyed due to a decrease in pH and oxygen levels and an increase in chloride ions, corrosion begins.

1.3.3.2 Pitting Corrosion

Pitting corrosion is also a localized type of corrosion that occurs in an implant even if it has small surface pits or holes [9]. In this type of corrosion, metal ions are released which then combine with chloride ions, initiating corrosion. Pitting corrosion is difficult to detect as the small pits are covered with corrosion products. It also results in roughening of the implant surfacedue to the formation of pits.

1.3.3.3 Galvanic Corrosion

Galvanic corrosion takes place when one metal corrodes before the other metal that is immersed in the same conductive solution. An example is a dental prosthesis where nickel and chromium may leach into the peri-implant tissues through the escape of saliva in the junction between the implant and its superstructure. This process affects the durability of the implant and may also lead to implant failure.

1.3.3.4 Electrochemical Corrosion

Electrochemical corrosion occurs when two different metals are in an electrolytic medium. In the case of biomedical implants, the body fluids act as an electrolyte and initiate coupled oxidation and reduction reactions resulting in metal deterioration as well as charge transfer via electrons. Most of the implanted metals, including cobalt-chromium, stainless steel, and titanium [10], lose electrons in solution and thus have a high potential to corrode.

1.3.3.5 Clinical Implication of Corrosion

Advances in technology have enabled the use of metallic [11] implants in the form of screws, pins, plates, artificial joints, and pacemakers; however, despite continuous development and research, no implant biomaterial is completely immune from corrosion. Therefore, research is ongoing in this field to develop implant biomaterials that are corrosion free as corrosion can result in roughening the surface and weakening the restoration, eventually leading to implant failure and inflammation caused by degraded products.

1.3.4 TRIBOLOGICAL PROPERTIES

The tribological behavior [12,13] of a biomaterial is one of the critical factors in determining its longevity when implanted in the body system. Tribological properties correlate with the frictional wear and lubrication of biomedical implants as this frictional behavior between the implant and bone interface affects the service period of the implant.

1.4 CLASSIFICATION OF BIOMATERIALS

1.4.1 CLASSIFICATION OF BIOMATERIALS BASED ON THEIR ORIGIN

Based on their origin, different types of biomaterials are categorized as synthetic (e.g., metals, polymers, and ceramics) or biological (e.g., collagen, silk, chitosan, and alginate. (Figure 1.2).

1.4.2 CLASSIFICATION OF BIOMATERIALS BASED ON THEIR BIOLOGICAL RESPONSE

1. **First generation (bioinert materials):** Once positioned in the human body, bioinert materials have minimal interaction or a chemical reaction with the surrounding tissues and structures [6].

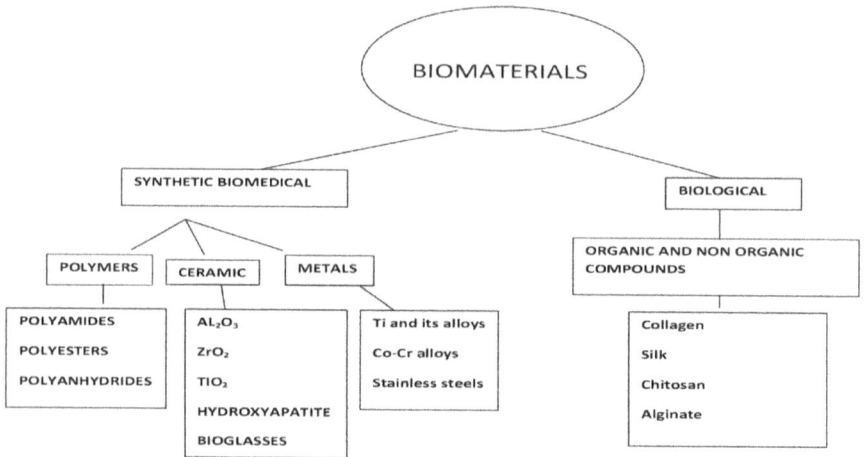

FIGURE 1.2   Classification of biomaterials based on their origin.

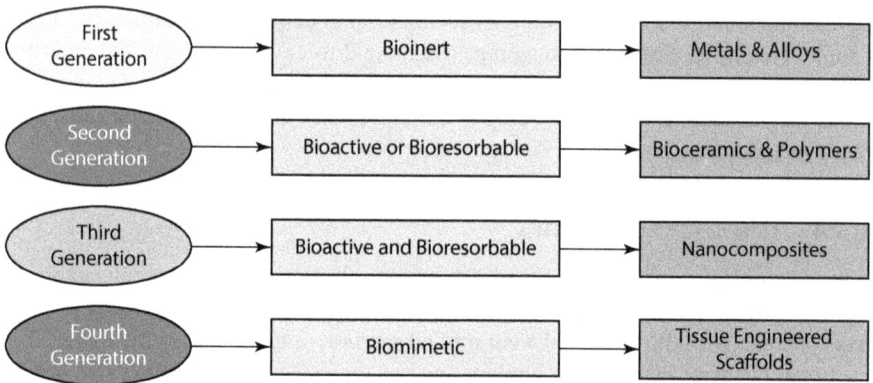

FIGURE 1.3   Classification of biomaterials based on their biological response.

2. **Second generation (bioactive materials or osteoconductive materials):** Also known as osteoconductive materials, they promote bone formation on their surface by acting as a scaffold through the process of ion exchange with tissue.

3. **Third generation (bioactive and bioresorbable):** Comprises nanocomposites that promote bone formation and undergo resorption to replace themselves as an autologous tissue.

4. **Fourth generation (biomimetic):** These materials are man-made synthetic materials that mimic themselves to simulate natural biological tissues. They are capable of adapting to dynamic environmental conditions (Figure 1.3).

1.5 TYPES OF BIOMATERIALS

1.5.1 METALLIC BIOMATERIALS

Metallic implants are the most commonly used biomaterials for joint replacement and are becoming increasingly important. The demand for bonehealingthrough internal fixation has led to a huge development in metallic implants and their clinical applications.

These metallic materials have excellent properties such as high strength, high fracture toughness, hardness, and biocompatibility.

The most commonly used metals for the fabrication of biomedical implants for orthopedic purposes are stainless steel, magnesium, titanium, and titanium alloys.

1. **Stainless steel:** Stainless steel is an alloy composed of iron (70%), chromium (16%–18%), nickel (10%–14%), and carbon (1%). 316L stainless steel is one of the most popular biomaterials to produce biomedical implants [15]. It provides a low-cost solution with good mechanical properties and biocompatibility and is therefore extensively used in the medical industry. Irrespective of its major advantages, stainless steel usage is becoming limited as it is prone to corrosion in a chloride environment and releases nickel and chromium ions that have detrimental effects on the human body. Stainless steel also has limited flexibility and thus is easily susceptible to deformation. Therefore, despite being a promising biomaterial for medical implants, stainless steel usage is better if used as a transitional or temporary implant [8]. Due to its propensity to corrode, permanent implants should not be fabricated from stainless steel. Future research should be carried out to improve the corrosion resistance as well as to enhance the ductility of stainless steel.

2. **Magnesium:** Magnesium-based biomedical implants are naturally biodegradable with a comparatively less elastic modulus similar to that of bone. It also possesses characteristics such as a high strength-to-weight ratio or specific strength, low density, and good biocompatibility and thus can be considered a promising substitute for permanent implants. However, magnesium undergoes a high rate of corrosion in an environment with a pH value of between 7.4 and 7.6 andthus its mechanical integrity declines prior to proper bone healing. In order to solve this, some techniques such as alloying and surface treatment are being used to control the rate of biodegradation.

3. **Titanium:** Titanium and its alloys are universally used in the fabrication of medical and dental implants due to their excellent biocompatibility and their property to form a stable oxide layer, thereby increasing corrosion resistance. Thus titanium and its alloys are used in numerous applications such as artificial joint replacement, bone fixation, neurosurgery, toe implants, false eye implants, and tooth replacement. Titanium also exhibits the excellent property of osseointegration, thereby increasing the structural and functional connection between living bone and the surface of a load-bearing artificial implant. Despite all of these advantages,implant failure

can occur in some cases with several reports highlighting allergic reactions such as urticaria, eczema, pain, necrosis, and bone loss.

1.5.2 BIOMEDICAL APPLICATION OF METALS

Metals are chosen for biomedical purposes according to the type of application. 316L stainless steel findswide application including the fabrication of cranial plates, dental [2] implants, stents, and catheters. However, forbiomedical uses require better corrosion and wear resistance, cobalt-chromium alloys are chosen [16].

Table 1.1 summarizes the various implant metals and their respective biomedical applications.

1.5.3 POLYMERS AND COMPOSITES

Polymeric biomaterials such as poly-alpha-hydroxy esters polylactic acidpolyglycolic acid (PGA), and their copolymers (PLGA) are extensively used in surgical and medical applications [17].

In general, when compared to other biomaterials, polymers possess low strength and modulus of elasticity but the magnitude of the elastic moduli is closer to that of soft tissues.

Polymeric composites possess better mechanical propertiescompared to purely polymeric materials. Their main advantage is that they enable the adhesion of human cells around them in a very organized way, which is essential for tissue regeneration.

Various polymeric composites such as polyethylene, polyether ether ketone, nano-hydroxyapatite, and ceramic biomaterials have come into use due to their superior properties.

TABLE 1.1
Summary of Various Implant Metals and Their Respective Biomedical Applications

Division	Application of implant	Type of metal used
Cardiovascular	Stent	316L stainless steel
	Artificial valve	Cobalt-chromium alloys
	Pacemaker	Titanium
Orthopedic	Bone fixation (plate, screws, pins)	316L stainless steel
	Artificial joints	Titanium, titanium alloys
	Spinal rods	Cobalt-chromium alloys
Dentistry	Dental implants	316L stainless steel
		Nitinol
Craniofacial	Plates and screws	316L stainless steel
	Orbit reconstruction	Titanium, titanium alloys
	Cranial plates	Cobalt-chromium alloys
Otorhinology	Artificial eardrum	316L stainless steel

Source: Langley and Dameron (2015) [14].

TABLE 1.2

Properties of Various Biomaterials

Properties	Polymers	Metals and alloys	Ceramics
Melting point	Low	Intermediate	High
Chemical stability	Poor	Good	Very high
Electrical conductivity	Very low	High	Very low but variable
Thermal conductivity	Very low to intermediate	High	Low
Properties and advantages	• Degradable • Inert • Ease of processing	• High strength and hardness	• Non-conductive and inert • Mimics biological properties of bone
Mechanical deformation	Very high	High (ductile)	Low (brittle)
Disadvantages	Thermally unstable Low strength	High wear and corrosion	Highly brittle High density

1.5.4 CERAMICS (BIOCERAMICS)

Ceramics are porous glass-based materials widely used in orthopedic and dental applications. Ceramics are inert materials with high hardness and high temperature and wear resistance; however, because of their low toughnessthey are highly fragile.

Zirconia and alumina are the most widely used ceramic oxides for biomedical purposes such as joint replacement and total hip replacement [18] as they possess excellent biocompatibility and wear resistance. Over time, research has been carried to improve ceramic properties through the development of nanocomposites and microsystems.

Nanocomposites show superior biocompatibility, osteoconductivity, and mechanical strength, making them a promising biomaterial for medical applications in the future.

The typical properties associated with different classes of biomaterials are summarized in Table 1.2 Ref. [19].

1.6 CONCLUSION

A wide range of biomaterials have been developedbased on their properties and biomedical applications. The correct selection of biomaterials based on clinical implications determinestheir clinical success and service duration. Conventional biomaterials such as stainless steel, magnesium, titanium, and cobalt-chromiumare widely used but because of their various limitations such as lower corrosion resistance, elastic modulus mismatch, and lower wear resistance, there is constant effort to improve their properties via surface coatings or by using nanotechnology to develop newer and better materials.

REFERENCES

1. Zwawi M. Recent advances in bio-medical implants; mechanical properties, surface modifications and applications. *Engineering Research Express.* 2022 Sep 1;4(3):032003.
2. Todros S, Todesco M, Bagno A. Biomaterials and their biomedical applications: From replacement to regeneration. *Processes.* 2021 Oct 29;9(11):1949.
3. Deepashree R, Devaki V, Kandhasamy B, Ajay R. Evolution of implant biomaterials: A literature review. *Journal of Indian Academy of Dental Specialist Researchers.* 2017;4(2):65.
4. Lane JM, Mait JE, Unnanuntana A, Hirsch BP, Shaffer AD, Shonuga OA. Materials in fracture fixation. *Comprehensive Biomaterials.* 2011:219–35.
5. Singh G, Singh H, Sidhu BS. In vitro corrosion investigations of plasma-sprayed hydroxyapatite and hydroxyapatite-calcium phosphate coatings on 316L SS. *Bulletin of Materials Science.* 2014 Oct;37(6):1519–28.
6. Moghadasi K, Mohd Isa MS, Ariffin MA, Mohd Jamil MZ, Raja S, Wu B, et al. A review on biomedical implant materials and the effect of friction stir based techniques on their mechanical and tribological properties. *Journal of Materials Research and Technology* [Internet]. 2022 Mar 1 [cited 2022 Apr 19];17:1054–121.
7. Henao J, Poblano-Salas C, Monsalve M, Corona-Castuera J, Barceinas-Sanchez O. Bio-active glass coatings manufactured by thermal spray: A status report. *Journal of Materials Research and Technology.* 2019 Sep 1;8(5):4965–84.
8. Gautam S, Bhatnagar D, Bansal D, Batra H, Goyal N. Recent advancements in nanomaterials for biomedical implants. *Biomedical Engineering Advances.* 2022 Jun;3:100029.
9. Saini M. Implant biomaterials: A comprehensive review. *World Journal of Clinical Cases.* 2015;3(1):52.
10. Jayaraj K, Pius A. Biocompatible coatings for metallic biomaterials. *Fundamental Biomaterials: Metals.* 2018:323–54.
11. Behera A. *Advanced Materials: An Introduction to Modern Material Science.* 1st ed Cham: Springer Nature 2021 Nov 22:439–67.
12. Shen G, Fang F, Kang C. Tribological performance of bioimplants: A comprehensive review. *Nanotechnology and Precision Engineering.* 2018 Jun 1;1(2):107–22.
13. Shen G, Zhang J, Culliton D, Melentiev R, Fang F. Tribological study on the surface modification of metal-on-polymer bioimplants. *Frontiers of Mechanical Engineering.* 2022 Jun;17:2–26.
14. Langley A, Dameron CT. Modern metal implant toxicity and anaesthesia. *Australian Anaesthesia.* 2015:57–65.
15. Love B. Metallic biomaterials. *In Biomaterials: A Systems Approach to Engineering Concepts.* Cambridge, MA: Academic Press. 2017:159–84.
16. Abraham AM, Venkatesan S. A review on application of biomaterials for medical and dental implants. *Proceedings of the Institution of Mechanical Engineers, Part L: Journal of Materials: Design and Applications.* 2022 Aug 29;237(2):249–73.
17. Oshida Y, Güven Y. *Biocompatible Coatings for Metallic Biomaterials.* 2015 Jan 1;287–343.
18. Vaiani L, Boccaccio A, Uva, AE, Palumbo G, Piccininni A, Guglielmi P, et al. Ceramic materials for biomedical applications: An overview on properties and fabrication processes. *Journal of Functional Biomaterials [Internet].* 2023 Mar 4 [cited 2023 Apr 21];14(3):146–6.
19. Wagner WR, Sakiyama-Elbert SE, Zhang G, Yaszemski MJ. *Biomaterials Science: An Introduction to Materials in Medicine* [Internet]. Academic Press. 2020 [cited 2023 Jul 3].

2 Synthesis and Characterization of Biocompatible Materials

Rajeev Singh, Avadesh K. Sharma,
and Ajay K. Sharma

2.1 INTRODUCTION

Generally, biomaterials are used to develop the structures and implants used to repair/replace the injured/lost biological structure, restoring its original shape and work. For example, the materials used for contact lenses, stents, orthodontic wire, and total hip replacements are biomaterials; however, the materials used for eyewear, hearing aids, and artificial prostheses are considered biomedical devices which are not included in the field of biomaterials [1,2].

Several biomaterials (e.g., metals, ceramics, and polymers) have been developed with suitable mechanical, biocompatible, and antibacterial properties. However, the requirement of these materials is based on the characteristics of the injured organ/tissue with a minimum toxic effect on the host environment [3,4]. Currently, 316L stainless steel, cobalt-chromium (Co-Cr), titanium (Ti), nickel-titanium (Ni-Ti), and magnesium (Mg) alloys are used as metallic biomaterials [5,6]. These materials are used for orthopedic implants due to their significant mechanical characteristics and reasonable biocompatibility [7,8]. Ceramics such as alumina (Al_2O_3), zirconia (ZrO_2), and other porous ceramics (e.g., hydroxyapatite, calcium phosphate, calcium pyrophosphate, and tri-calcium phosphate) are also used as biomaterials. These ceramics are used because of their high toughness, wear resistance, and biocompatibility [9]. Polymers such as polyethylene (PE), polypropylene (PPE), and poly-lactic acid (PLA) are promising biomaterials. Today, polymeric biomaterials are more favorable owing to their low weight, high-impact strength, and better biocompatibility [10].

The properties of biomaterials are based on the homogeneous distribution of matrix particles. The homogeneous distribution of particles provides grain refinement and a defect-free microstructure that lead to improved mechanical and biological properties owing to required application. A defect-free microstructure and optimum properties can be achieved by selecting a suitable method for the fabrication of nanocomposites [11,12].

DOI: 10.1201/9781003375098-2

2.2 SYNTHESIS OF BIOMATERIALS

The synthesis of biomaterials mainly depends on two approaches: the production of fine powder materials or the production of bulk materials. The synthesis approach in which fine powders are produced is categorized as a top-down approach. The approach related to the synthesis of bulk biomaterials is categorized as a bottom-up approach (Figure 2.1). The top-down approach is related to the reduction in material size from top (bulk) to bottom (small). In this approach, macroscopic structures are used for the production of biomaterials. The biomaterials are synthesized by extracting particles through etching crystal planes from the bulk material [13–15]. In the bottom-up approach, the biomaterials are extracted by adding atoms to each other, resulting in crystal planes stacked on top of each other. Chemical and physical forces are used during the self-assembly of atoms to transform them into stable microstructures. Similarly, synthesis techniques are classified as per the type of biomaterial (Figure 2.2). These techniques are also categorized as solid- and liquid-state processes. Biomaterials are synthesized below melting temperature (e.g.,

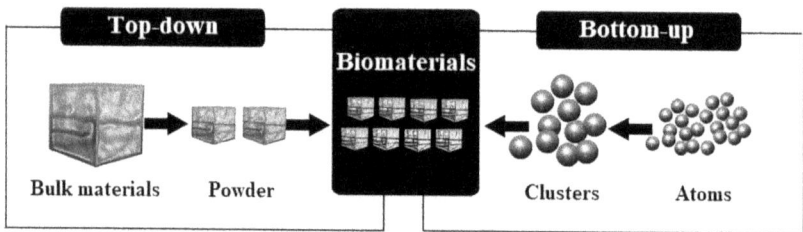

FIGURE 2.1   Two different approaches for the synthesis of biomaterials.

Synthesis techniques of biomaterials		
Metallic biomaterials	Ceramic biomaterials	Polymeric biomaterials
Infiltration	Powder metallurgy	Melt blending
Powder metallurgy	Electrophoretic deposition (EPD)	Sol-gel
Spray pyrolysis	Sol-gel	
Electrophoretic deposition (EPD)		
Sol-gel		

FIGURE 2.2   Classification of synthesis techniques used for different biomaterials.

powder metallurgy [PM] processes) in solid-state processes, whereas in liquid-state processes the biomaterials are synthesized by the melted-state process or above the melting point in the infiltration process [16]. The techniques used for the synthesis of different types of biomaterials are discussed next.

2.2.1 INFILTRATION PROCESS

Infiltration is a technique in which biomaterials are developed by blending ceramic particles or fibers in a liquid-state metal matrix. It is one of the best approaches for synthesizing a metallic biomaterial because the uniform dispersion of the matrix particles leads to sufficient mechanical strength [17]. In this process, the melted metal fills the porous matrix with or without the application of forces under a controlled atmosphere to achieve better wetting conditions (Figure 2.3) [18]. In pressure-less infiltration, no external force is needed to complete the operation. In forced infiltration, external pressure is used to accelerate the penetration of the reinforcement into the molten metal matrix [19]. A biomaterial developed using the infiltration process shows a uniform distribution of phases and a dense microstructure and is successfully used in biomedical applications [20–22].

2.2.2 POWDER METALLURGY PROCESS

PM is a technique in which metallic or ceramic powders are mixed and shaped under the application of pressure and temperature [23]. The process involves three basic steps: mixing, compacting, and sintering. A specific composition of powder materials is mixed in a ball mill, a process known as mechanical alloying. The homogeneous mixing of powder is totally based on a parameter such as the milling time and speed [24]. Compaction is a process that is carried out to form a green compact by pressing the powder mixture in a closed die under high pressure. The powder particles are attached in a specific shape due to the activated green strength between the particles. Finally, sintering is performed at a specific temperature to improve the microstructure and mechanical strength of a green compact. The sintering process is carried out in a controlled atmosphere to avoid oxidation. Generally, researchers use vacuum, spark plasma, and microwave sintering to develop biomaterials [25]. Powder metallurgy is a cost-effective method that is used to fabricate biomaterials

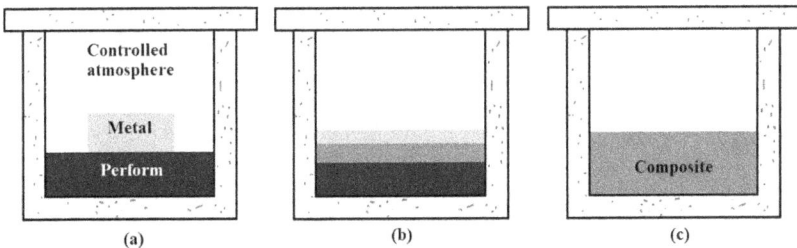

FIGURE 2.3 Schematic diagram of pressure-less infiltration.

Mixing

Raw powders

Power
supply

Compaction

Final composites

Gas out

Gas in

Sintering   Green
                   compact

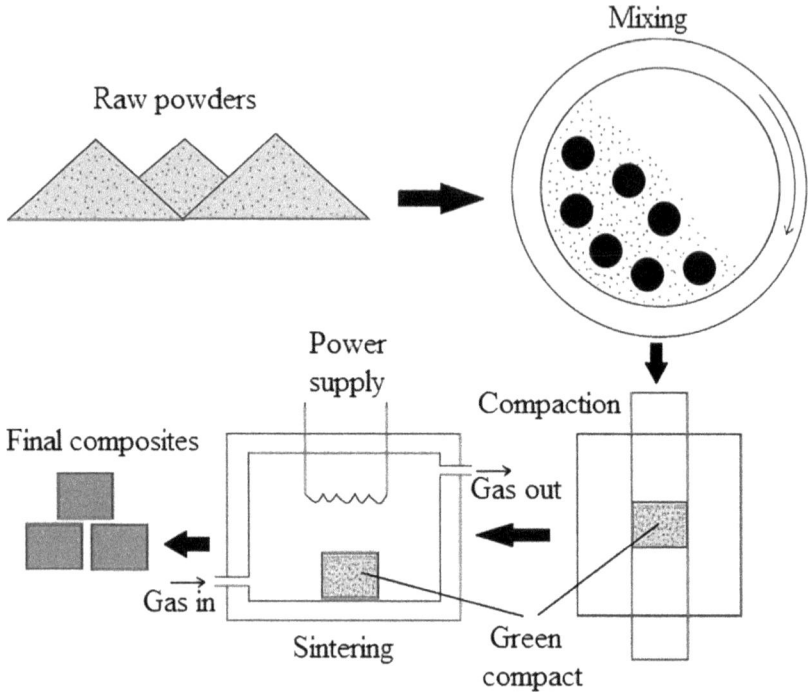

FIGURE 2.4   Schematic diagrams of the powder metallurgy process.

with a fine-grained and homogeneous microstructure. Any type of hard material such as metals and ceramics can be produced using this process [26] (Figure 2.4).

2.2.3   SPRAY PYROLYSIS

In this technique, a precursor solution is atomized in small drops and sprayed on a heated substrate to create a thin coating. This process contains a precursor solution, atomizer, substrate heater, and temperature controller. In this process, an inorganic precursor is dissolved in a solvent to produce a liquid source. Using an ultrasonic atomizer, the liquid source is atomized in the form of a mist. The mist is carried by a gas into a heated chamber and vaporized into small droplets. These small droplets are decomposed in the respective oxide materials and further deposited on the metallic materials [27]. It is a continuous and single-step technique that is used to produce biomaterials with particles of a uniform shape and size. Ultrafine particles in non-agglomerated form can also be obtained with this method.

2.2.4   ELECTROPHORETIC DEPOSITION

Generally, metallic and ceramic biomaterials that are used for the fabrication of medical implants have a bioinert nature. The bioinert nature of these biomaterials

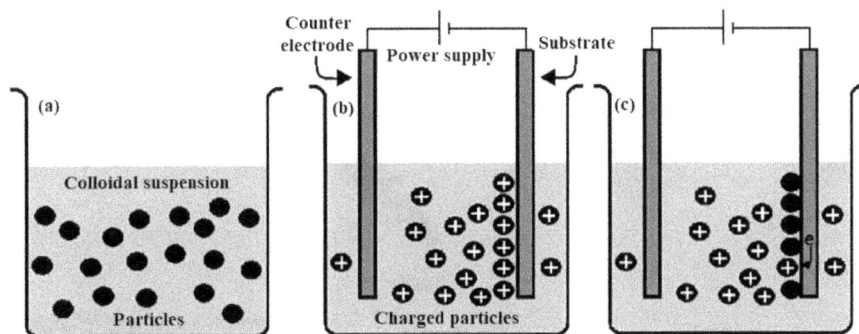

FIGURE 2.5 Schematic diagrams of EPD showing (a) colloidal suspension, (b) charged particles, and (c) deposition.

can be enhanced by forming a bioactive coating on the implant surfaces. This bioactive coating can be easily formed by electrophoretic deposition (EPD) [28]. EPD is a colloidal deposition technique controlled by a DC electric field [29,30]. In this process, the colloidal powder particles are dispersed and charged in a solvent suspension (Figure 2.5a,b). These charged particles are moved electrophoretically in the suspension and accumulate on the opposite charge electrode (conducting surface) under a DC electric field (Figure 2.5c) [31]. EPD shows several interesting features such as easy processing and control, cost-effectiveness, low processing temperature, higher production rate, and the ability to form complex shapes and structures as per the requirements of biomedical applications. It also offers the alteration of the morphology of a deposited coating on the substrate by simple adjustment of the EPD parameters (e.g., deposition time and applied potential).

2.2.5 SOL-GEL PROCESS

This is a wet-type chemical process for synthesizing nanostructured biomaterials such as powders, coatings, and composites [32]. In this process, the biomaterials are produced by mixing the colloidal suspension (sol) and the gelation (gel) to develop a mixed liquid–phase system. The raw materials are evenly mixed in the liquid state for hydrolysis as well as the polycondensation process to create a transparent/stable sol. The colloidal particles gradually accumulate to form a gel of three-dimensional structures after aging of the sols. After drying and sintering, the biomaterials are prepared with the required structures [33]. Sol-gel is an easy and convenient technique to synthesize biomaterials with better control of the product's chemical composition [34,35]. Pazarçeviren et al. [36] synthesized bismuth-doped 45S5 nano-bioactive bioglass (nBG) using the sol-gel process. The results show that the synthesis of Bi-doped 45S5 nano-bioactive bioglass was successfully carried out using the sol-gel process. Hou et al. [37] synthesized a silk and silica (SiO_2)-based hybrid biomaterial using the sol-gel process. The nanoparticles were successfully

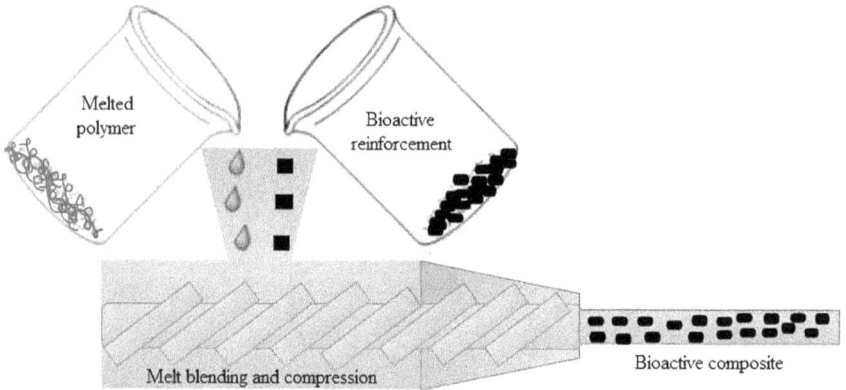

FIGURE 2.6   Synthesis of biomaterials by melt blending.

synthesized using the sol-gel technique with a heterogeneous structure and a higher purity due to covalent cross-linking.

2.2.6   MELT BLENDING

Melt blending is the simplest technique for the development of polymer-based biomaterials. Several types of polymeric biomaterials, such as styrene butadiene rubber, are synthesized with this technique [38]. Using this technique, the elastomeric polymer is melted and mixed with the required amount of reinforcement particles. The polymeric mixture is compressed to the desired shape in an extruder (Figure 2.6) [39]. The process is performed in an inert gas environment using argon or nitrogen. Melt blending is a synthesis technique that reduces the possibility of material contamination. However, the possibility of polymer degradation during melting and the high cost of equipment restrict its application in biomedical applications. Additionally, the synthesis of small quantities of materials is quite difficult with this technique [40,41].

2.3   CHARACTERIZATION OF BIOMATERIALS

2.3.1   MICROSTRUCTURAL CHARACTERIZATION

Initially, the biomaterials are prepared for microstructural characterization using grinding, polishing, lapping, and etching operations. Further, the microstructural characterization of biomaterials is performed by scanning electron microscopy (SEM). The corresponding elemental details of the biomaterials are examined by energy-dispersive x-ray spectroscopy (EDS). The phases formed during the synthesis of biomaterials are identified by x-ray diffraction (XRD). The operations and techniques used for the microstructural characterization of biomaterials are discussed below.

2.3.1.1 SEM

An SEM is an electron-based microscope used to observe very small morphological or microstructural details on the surface of objects. A schematic of an SEM is shown in Figure 2.7. The object is scanned by negatively charged electrons that are discharged by a field emission source and accelerated in a large electrical field. In this technique, electronic lenses are used that focus the primary electrons to bombard a narrow beam on the surface in a high vacuum column. Consequently, secondary electrons are reflected at a specific speed and angle onto the surface microstructure. These electrons are detected by a detector, which generates an electrical signal. This signal is amplified and transformed into an image that is observed on a monitor. In the present work, the microstructures of sintered samples are examined at an accelerated voltage of 15 kV with the Everhart–Thornley detector (ETD) in field-free lens mode. Images up to 10,000× magnification are captured with a 10 second dwell time.

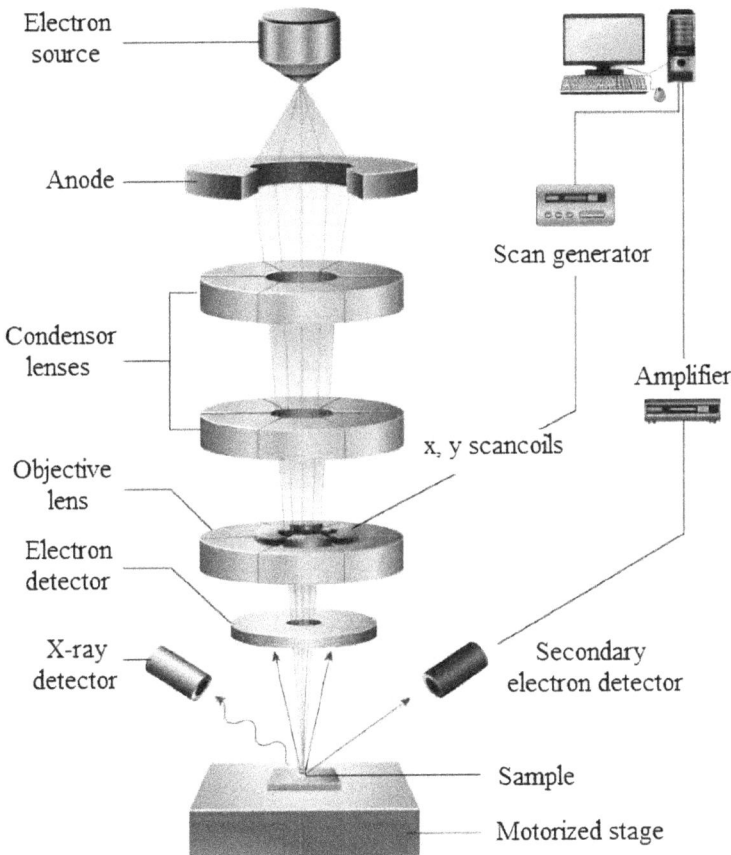

FIGURE 2.7 Schematic of an SEM.

2.3.1.2 EDS

EDS is a standard technique alongside the SEM that identifies and quantifies the elements present in a given material. In EDS, a high-intensity beam of charged electrons is directed to the object and characteristic x-rays are developed. The beam of the electron excites the inner cell electron and then ejects it to produce an electron hole. The hole is filled with the electrons of the outer shell upon release of x-rays from the sample. The energy and amount of x-rays are measured by an energy-dispersive spectrometer which identifies and quantifies the elements present in the sample.

2.3.1.3 XRD

XRD primarily examines the phases of a crystalline material. XRD is a combination of an x-ray tube, a detector, and a sample holder. X-rays are generated in a cathode ray tube (CRT) by filament heating for the production of electrons. By applying a voltage, these electrons are accelerated and bombarded on the target. When the electrons contain sufficient energy to escape from the inner core, they produce characteristic x-ray spectra. These x-rays are collected and directed toward the target. Simultaneously, the detector detects the intensity of the emitted x-rays due to the rotation of the target and detector. When the geometry of the emitted x-rays satisfies Bragg's law, these x-ray signals are collected and transformed into peak intensity and count rate by the detector.

2.3.2 DENSIFICATION CHARACTERIZATION

2.3.2.1 Density and Porosity of Biomaterials

The density of biomaterials is generally calculated using Archimedes' principle. Further, the relative density ($\rho_{relative}$) and porosity (p) of biomaterial samples are measured using Eqs. (2.1) and (2.2), respectively. The porosity of the composites is also analyzed by software such as ImageJ (ASTM e562 and e1245).

$$\rho_{relative} = \frac{\rho_{sintering}}{\rho_{theoretical}} \times 100 \tag{2.1}$$

$$p = \left\{ 1 - \left(\frac{\rho_{sintering}}{\rho_{theoretical}} \right) \right\} \times 100 \tag{2.2}$$

2.3.2.2 Water Absorption Test

A water absorption test is carried out to identify the amount of interconnected pores in the samples of biomaterials. The samples are weighed before immersion and following a certain time period of immersion in deionized water. Finally, the water absorption is identified using Eq. (2.3) which represents the amount of interconnected pores in the samples.

$$W_{absorption} = \left\{ \frac{\left(m_{final} - m_{initial} \right)}{m_{initial}} \right\} \times 100 \tag{2.3}$$

where $w_{absorption}$ is the percentage of water absorbed, and $m_{initial}$ and m_{final} are the mass of the sample before and after immersion.

2.3.3 MECHANICAL CHARACTERIZATION

2.3.3.1 Compression Test

Compression testing is performed for the mechanical characterization of biomaterials on the basis of properties such as the elastic modulus and yield strength. Generally, the test is performed on a universal testing machine (UTM), as shown in Figure 2.8a. A UTM can perform many mechanical tests on materials such as tensile, compression, bending, and tear. The compression test is performed according to ISO 13314 at room temperature. In this test, a solid sample is compressed by the action of opposing forces that push the sample from opposite sides. The sample is reduced in length and expands perpendicular to the load axis. When the sample is fully compressed and it is plastically deformed or fractured, the sensors record the corresponding load, stress, and strain. The results are displayed on a computer monitor.

2.3.3.2 Hardness Test

The hardness of biomaterials is identified by a Vickers hardness tester, as shown in Figure 2.8b. The tester measures at a variable loading speed with a display resolution of up to 0.1 HV. During Vickers hardness testing, a diamond indenter indents a sample surface under an applied load for a specified hold time. Following indentation, the indent size is measured using an attached microscope. Finally, the hardness is measured as per the mean stress applied by the indenter.

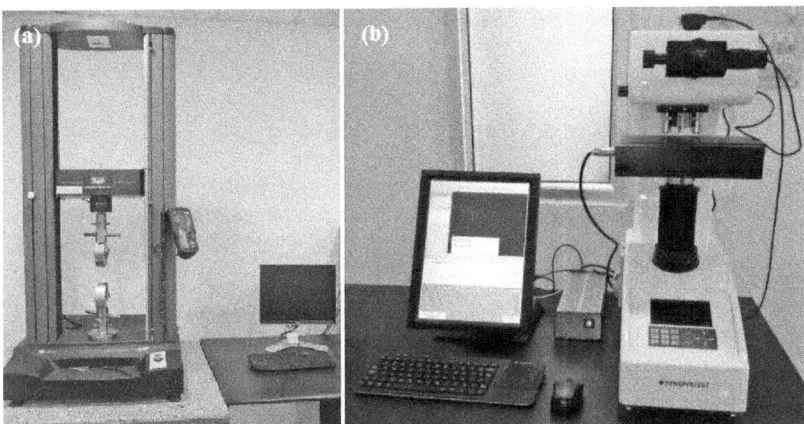

FIGURE 2.8 (a) UTM (Instron-5967, USA) and (b) Vickers hardness tester (Innovatest, the Netherlands).

2.3.4 BIOLOGICAL CHARACTERIZATION

2.3.4.1 Immersion Test

An immersion test is carried out in a simulated body fluid (SBF) to identify the release of ions from the biomaterials surface. The SBF was developed according to the method given by Kokubo and Takadama [42]. The polished samples are immersed in SBF for a specific time period under a physiological environment (i.e., 37°C, pH 7.4). Following the immersion test, the amount of metallic ions accumulated in the SBF is analyzed by an atomic absorber spectrometer (AAS). An AAS is a simple, low-cost, and accurate technique for determining the elements present in a solution. It detects elements by using the characteristic wavelengths of electromagnetic radiation emitted by a light source. Each element in the solution absorbs different wavelengths of radiation, and these absorbances are measured against predefined standards. Initially, the solution is atomized to emit and record its characteristic wavelengths. Then, atoms absorb a specific energy and corresponding electrons move to another energy level due to excitation. This energy relates to a specific wavelength that shows the properties of the element. Finally, the specific elements are detected with their concentrations on the basis of the intensity of these wavelengths.

2.3.4.2 In Vitro Bioactivity

The bioactivity (osseointegration) of biomaterials is also examined by immersion testing [43]. In this test, the required amount of SBF is calculated using Eq. (2.4):

$$\frac{S}{V} = 0.05 \text{ cm}^{-1} \tag{2.4}$$

where S is the open sample area and V is the required amount of SBF. On completion of the immersion time, the samples are removed from the SBF, washed, and dried. The development of an apatite layer (i.e., bioactivity) on the biomaterials' surface is examined using an SEM, and the quantification of deposited Ca-P ions is identified by EDS.

2.4 CONCLUSION

This chapter deals with the synthesis techniques and characterization of biomaterials. Several types of biomaterials (i.e., metals, ceramics, and polymers) are synthesized with the techniques given in this chapter. The selection of the synthesis technique is very important for achieving suitable properties as per applications. It is observed that metallic and ceramic biomaterials are synthesized by the powder metallurgy process due to the higher melting temperatures of matrix materials; however, polymer-type biomaterials are synthesized by the melt blending process. These processes provide uniform dispersion of matrix particles leading to improved performance in biomedical applications. The characterization methods used for identifying the performance of biomaterials are also described in this chapter. The microstructural characterization defines the microstructure, morphology, and phases of biomaterials. Similarly, the mechanical and biological characterizations report the properties of biomaterials, such as strength, elastic modulus, and bioactivity.

REFERENCES

1. Orive, G., Ali, O. A., Anitua, E., Pedraz, J. L., Emerich, D. F. (2010). Biomaterial-based technologies for brain anti-cancer therapeutics and imaging 1806(1), 96–107. doi:10.1016/j.bbcan.2010.04.001
2. Pei, Y., Zhang, L., Mao, X., Liu, Z., Cui, W., Sun, X., Zhang, Y. (2020). Biomaterial scaffolds for improving vascularization during skin flap regeneration. *Chinese Journal of Plastic and Reconstructive Surgery*, 2(2), 109–119. doi:10.1016/s2096-6911(21)00021-2
3. Ullah, S., Chen, X. (2020). Fabrication, applications and challenges of natural biomaterials in tissue engineering. *Applied Materials Today*, 20, 100656. doi:10.1016/j.apmt.2020.100656
4. Zaokari, Y., Persaud, A., Ibrahim, A. (2020). Biomaterials for adhesion in orthopedic applications: A review. *Engineered Regeneration*, 1, 51–63. doi:10.1016/j.engreg.2020.07.002
5. Dogra, V., Kaur, G., Jindal, S. et al. (2019). Bactericidal effects of metallosurfactants based cobalt oxide/hydroxide nanoparticles against Staphylococcus aureus. *Science of the Total Environment*, 681, 350–364. doi:10.1016/j.scitotenv.2019.05.078
6. Jenko, M., Gorenšek, M., Godec, M. et al. (2018). Chemistry and microstructure of metallic biomaterials for hip and knee endoprostheses. *Applied Surface Science*, 427, 584–593. doi:10.1016/j.apsusc.2017.08.007
7. Stewart, C., Akhavan, B., Wise, S. G. et al. (2019). A review of biomimetic surface functionalization for bone-integrating orthopedic implants: Mechanisms, current approaches, and future directions. *Progress in Materials Science*, 106, 100588. doi:10.1016/j.pmatsci.2019.100588
8. Geetha, M., Singh, A. K., Asokamani, R. et al. (2009). Ti based biomaterials, the ultimate choice for orthopaedic implants – A review. *Progress in Materials Science*, 54(3), 397–425. doi:10.1016/j.pmatsci.2008.06.004
9. Navarro, M., Michiardi, A., Castaño, O., Planell, J. A. (2008). Biomaterials in orthopaedics. *Journal of the Royal Society, Interface*, 5(27), 1137–1158. doi:10.1098/rsif.2008.0151
10. Yang, R., Chen, F., Guo, J., Zhou, D., Luan, S. (2020). Recent advances in polymeric biomaterials-based gene delivery for cartilage repair. *Bioactive Materials*, 5(4), 990–1003. doi:10.1016/j.bioactmat.2020.06.004
11. Nishida, Y. (2013). *Introduction to Metal Matrix Composites: Fabrication and Recycling*. Springer Science & Business Media. doi:10.1007/978-4-431-54237-7
12. Camargo, P. H. C., Satyanarayana, K. G., Wypych, F. (2009). Nanocomposites: Synthesis, structure, properties and new application opportunities. *Materials Research*, 12(1), 1–39.
13. D'Costa, G., Pisal, D. S., Rane, A. V. (2012). Report on synthesis of nanoparticles and functionalization: Co precipitation method.
14. Chikan, V., McLaurin, E. J. (2016). Rapid nanoparticle synthesis by magnetic and microwave heating. *Nanomaterials*, 6(5), 85.
15. Kountouras, D. T., Vogiatzis, C. A., Stergioudi, F., Tsouknidas, A., Skolianos, S. M. (2014). Optimizing ceramic preform properties for liquid metal infiltration. *Journal of Materials Engineering and Performance*, 23(6), 2015–2019.
16. Koli, D. K., Agnihotri, G., Purohit, R. (2014). A review on properties, behaviour and processing methods for Al– Nano Al_2O_3 composites. *Procedia Materials Science*, 6, 567–589.

17. Kountouras, D. T., Vogiatzis, C. A., Stergioudi, F., Tsouknidas, A., Skolianos, S. M. (2014). Optimizing ceramic preform properties for liquid metal infiltration. *Journal of Materials Engineering and Performance*, 23(6), 2015–2019. doi:10.1007/s11665-014-1022-8

18. Xiong, B., Xu, Z., Yan, Q., Lu, B., Cai, C. (2011). Effects of SiC volume fraction and aluminum particulate size on interfacial reactions in SiC nanoparticulate reinforced aluminum matrix composites. *Journal of Alloys and Compounds*, 509(4), 1187–1191. doi:10.1016/j.jallcom.2010.09.171

19. Gan, Y. X. (2012). Structural assessment of nanocomposites. *Micron*, 43(7), 782–817. doi:10.1016/j.micron.2012.02.004

20. Sree Manu, K. M., Ajay Raag, L., Rajan, T. P. D., Gupta, M., Pai, B. C. (2016). Liquid metal infiltration processing of metallic composites: A critical review. *Metallurgical and Materials Transactions. Part B*, 47(5), 2799–2819.doi:10.1007/s11663-016-0751-5

21. Pfeiffer, S., Lorenz, H., Fu, Z., Fey, T., Greil, P., Travitzky, N. (2018). $Al_2O_3/Cu-O$ composites fabricated by pressureless infiltration of paper-derived Al_2O_3 porous preforms. *Ceramics International*, 44(17), 20835–20840. doi:10.1016/j.ceramint.2018.08.087

22. Cramer, C. L., Preston, A., Elliott, A. M., Lowden, R. A. (2019). Highly dense, inexpensive composites via melt infiltration of Ni into WC/Fe preforms. *International Journal of Refractory Metals and Hard Materials*, 82, 255–258.doi:10.1016/j.ijrmhm.2019.04.019

23. Suryanarayana, C., Al-Aqeeli, N. (2013). Mechanically alloyed nanocomposites. *Progress in Materials Science*, 58(4), 383–502.

24. Singh, R., Sharma, A. K., Sharma, A. K. (2021). Nickel-titanium based nanocomposites for orthopedic applications: The effects of Reinforcements. *Digest Journal of Nanomaterials and Biostructures*, 16(4), 1501–1518.

25. Jayasathyakawin, S., Ravichandran, M., Baskar, N., Anand Chairman, C., Balasundaram, R. (2019). Magnesium matrix composite for biomedical applications through powder metallurgy – Review. *Materials Today: Proceedings*, 27, 736–41.

26. Abdullah, R., Adzali, N. M. S., Daud, Z. C. (2016). Bioactivity of a bio-composite fabricated from CoCrMo/Bioactive glass by powder metallurgy method for biomedical application. *Procedia Chemistry*, 19, 566–570.

27. Choa, Y.-H., Yang, J.-K., Kim, B.-H., Jeong, Y.-K., Lee, J.-S., Nakayama, T., Sekino, T., Niihara, K. (2003). Preparation and characterization of metal/ceramic nanoporous nanocomposite powders, 266(1–2), 12–19. doi:10.1016/s0304-8853(03)00450-5

28. Besra, L., Liu, M. (2007). A review on fundamentals and applications of electrophoretic deposition (EPD). *Progress in Materials Science*, 52(1), 1–61. doi:10.1016/j.pmatsci.2006.07.001

29. Say, Y., Aksakal, B. (2020). Enhanced corrosion properties of biological NiTi alloy by hydroxyapatite and bioglass based biocomposite coatings. *Journal of Materials Research and Technology*, 9(2), 1742–1749. doi:10.1016/j.jmrt.2019.12.005

30. Horandghadim, N., Khalil-Allafi, J., Kaçar, E., Urgen, M. (2019). Biomechanical compatibility and electrochemical stability of HA/Ta2O5 nanocomposite coating produced by electrophoretic deposition on superelastic NiTi alloy. *Journal of Alloys and Compounds*, 799, 193–204. doi:10.1016/j.jallcom.2019.05.166

31. Aliofkhazraei, M., Makhlouf, A. S. H. (2016). Electrophoretic Deposition (EPD): Fundamentals and applications from nano- to micro-scale structures. *Handbook of Nanoelectrochemistry*, 1–27. doi:10.1007/978-3-319-15207-3_7-1

32. Rane, A. V., Kanny, K., Abitha, V. K., Thomas, S. Methods for synthesis of nanoparticles and fabrication of nanocomposites. In S.M. Bhagyaraj, O.S. Oluwafemi, N. Kalarikkal, S. Thomas (eds.), *Synthesis of Inorganic Nanomaterials*, Elsevier (2018), 121–139.

33. Singh, L. P., Bhattacharyya, S. K., Kumar, R., Mishra, G., Sharma, U., Singh, G., Ahalawat, S. (2014). Sol-gel processing of silica nanoparticles and their applications. *Advances in Colloid and Interface Science*, 214, 17–37.
34. Owens, G. J., Singh, R. K., Foroutan, F., Alqaysi, M., Han, C.-M., Mahapatra, C., Knowles, J. C. (2016). Sol–gel based materials for biomedical applications. *Progress in Materials Science*, 77, 1–79.
35. Almeida, V. O., Balzaretti, N. M., Costa, T. M. H., Gallas, M. R. (2015). Enhanced mechanical properties in ZrO2 multi-walled carbon nanotube nanocomposites produced by sol–gel and high-pressure. *Nano-Structures and Nano-Objects*, 4, 1–8.
36. Pazarçeviren, A. E., Tahmasebifar, A., Tezcaner, A., Keskin, D., Evis, Z. (2018). Investigation of bismuth doped bioglass/graphene oxide nanocomposites for bone tissue engineering. *Ceramics International*, 44(4), 3791–3799.
37. Hou, A., Chen, H. (2010). Preparation and characterization of silk/silica hybrid biomaterials by sol–gel crosslinking process. *Materials Science and Engineering. Part B*, 167(2), 124–128.
38. Ivanoska-Dacikj, A., Bogoeva-Gaceva, G. Fabrication methods of carbon-based rubber nanocomposites. In Srinivasarao Yaragalla, Raghvendra Kumar Mishra, Sabu Thomas, Nandakumar Kalarikkal, Hanna J. Maria (eds.), *Carbon-Based Nanofillers and Their Rubber Nanocomposites*, Elsevier (2019), 27–47. doi:10.1016/B978-0-12-817342-8.00002-0
39. Gangarapu, S., Sunku, K., Babu, P. S., Sudarsanam, P. Fabrication of polymer-graphene nanocomposites. In C. Hussain, S. Thomas (eds.), *Handbook of Polymer and Ceramic Nanotechnology*, Springer, Cham (2020). doi:10.1007/978-3-030-10614-0_31-1
40. Cruz, S. M., Viana, J. C. (2014). Melt blending and characterization of carbon nanoparticles-filled thermoplastic polyurethane elastomers. *Journal of Elastomers and Plastics*, 1–19. doi:10.1177/0095244314534097
41. Toh, H. W., Toong, D. W. Y., Ng, J. C. K., Ow, V., Lu, S., Tan, L. P., Wong, P. E. H., Venkatraman, S., Huang, Y., Ang, H. Y. (2021). Polymer blends and polymer composites for cardiovascular implants. *European Polymer Journal*, 146, 110249.
42. Kokubo, T., Takadama, H. (2006). How useful is SBF in predicting in vivo bone bioactivity? *Biomaterials*, 27(15), 2907–2915. doi:10.1016/j.biomaterials.2006.01.017
43. Maleki-Ghaleh, H., Khalil-Allafi, J. (2019). Characterization, mechanical and in vitro biological behavior of hydroxyapatite-titanium-carbon nanotube composite coatings deposited on NiTi alloy by electrophoretic deposition. *Surface and Coatings Technology*. doi:10.1016/j.surfcoat.2019.02.029

3 Recent Advancements in Materials

Nanocomposites

*S.S.P.M. Sharma B, Anupma Agarwal,
Premanand S. Chauhan,
Rachna Zirath, and Amit Khemka*

3.1 INTRODUCTION

Composites are a combination of two or more constituent materials, each possessing distinct physical or chemical properties. As a result of blending process, a remarkable transformation occurs, giving rise to a composite material with entirely novel and unique characteristics that transcend those of its individual components [1]. The aerospace, transportation, military, sports equipment, and electronic devices industries all make substantial use of human-engineered composites due to their outstanding functional and mechanical qualities [2].

Nanocomposites, in particular, have attracted significant attention from the research community due to their exceptional properties, which are tailored by incorporating nanoscale fillers such as 0D particles (fullerene), 1D nanofibers (carbon nanotubes [CNT]), and 2D nanosheets (montmorillonite, graphene, and boron nitride). Among these fillers, 2D nanosheets exhibit superior properties compared to 0D or 1D nanofillers because they possess a much higher specific area [2,3].

The recent progress in producing nanostructured materials with unique properties has sparked a wave of research aimed at designing macroscopic engineering materials with multiple functions through nanoscale structuring [4–6]. In this context, the development of nanocomposites has emerged as a rapidly advancing field within composites research, fueled by the widespread interest in nanotechnology. Researchers are exploring the incorporation of nanoparticles, nanofibers, and nanosheets into composites to enhance their performance and enable new functionalities. This passion for nanotechnology has propelled the exploration of nanocomposites and holds promise for creating innovative materials with enhanced properties [7–9].

Nanotechnology can be broadly defined as a field encompassing the creation, processing, characterization, and utilization of materials, devices, and systems that exhibit novel and significantly enhanced physical, chemical, and biological properties, functions, phenomena, and processes due to their nanoscale size, typically ranging from 0.1 to 100 nm [10,11].

DOI: 10.1201/9781003375098-3

The scope of nanotechnology includes various aspects such as nanobiotechnology, nanosystems, nanoelectronics, and nanostructured materials [12,13]. Within this broad framework, nanocomposites play a significant role. Nanocomposites refer to composite materials where nanoscale fillers, such as nanoparticles, nanofibers, or nanosheets, are incorporated into a matrix material [14]. These nanoscale fillers impart unique properties and functionalities to the composites, thereby enhancing their overall performance. Nanocomposites are an important area of research and application within the field of nanotechnology, offering great potential for the development of advanced materials with tailored properties [13,14].

A major morphological aspect in the structure–property connection of nanocomposites is the surface area-to-volume ratio of reinforcing components. From meters (completed woven composite parts) to micrometers (fiber diameter) to sub-micrometers (fiber/matrix interphase) to nanometers (nanotube diameter), the surface area of a material grows in proportion to its volume [15].

New methods for processing, characterizing, analyzing, and modeling this class of composite materials are made possible by their increased surface area. Their higher surface area allows for enhanced interfacial interactions between the reinforcement materials and the matrix, leading to improved mechanical, electrical, thermal, and other functional properties of nanocomposites [16,17].

By understanding and manipulating the surface area-to-volume ratio at different length scales, researchers can tailor the properties of nanocomposites to meet specific requirements. This opens up exciting possibilities for designing and developing advanced composite materials with enhanced performance and functionality for a wide range of applications [18,19].

3.2   COMPOSITES REINFORCED WITH NANOPARTICLES

"Particulate composites" are widely used because they are reinforced with microscopic particles of several different materials. Typically, these particles are added to a matrix in an effort to increase its yield strength and elastic modulus. However, it has been demonstrated that new material qualities can be attained by decreasing the particle size to the nanoscale [19,20].

Increased optical transparency and altered fracture toughness are two ways in which nanocomposites improve upon conventional composites. Particulate composites are often opaque because the presence of micron-sized particles scatters light, making the transparent matrix material opaque. However, studies by Naganuma and Kagawa on SiO_2/epoxy composites showed that particle size gains in visible light transmission might be achieved by lowering the particle size [21,22].

Similarly, Singh et al. investigated the impact of including aluminum particles with diameters of 20, 3.5, and 100 nm on the fracture toughness of polyester resin. Their study demonstrated that the incorporation of nanoscale aluminum particles resulted in notable changes to the fracture toughness of the composite material. These findings illustrate the potential of nanoscale reinforcement in tailoring the optical and mechanical properties of composites, offering new possibilities for a range of applications [23].

Lopez and colleagues conducted a study on vinyl ester composites, investigating the effect of incorporating alumina particles with diameters of 40 nm, 1 μm, and 3 μm at weight fractions of 1, 2, and 3 wt%. They observed a monotonous increase in the composite modulus with increasing particle weight fraction for all particle sizes. However, due to non-uniform particle size distribution and particle aggregation, the composite strengths were found to be lower than those of the neat resin [24].

Thompson et al. encountered similar challenges in their study on metal oxide/polyimide nanocomposite films. They incorporated antimony tin oxide, indium tin oxide, and yttrium oxide nanoparticles into two space-durable polyimides: TOR-NC and LaRC TMCP-2. The addition of nanoscale particles led to improved stiffness but equivalent or lower strengths and elongation, as well as lower dynamic stiffness. However, achieving a uniform dispersion of metal oxides at the nanoscale within the composite materials proved to be challenging for the researchers [24].

These findings highlight the complexities associated with processing and achieving uniform dispersion of nanoscale additives within composites, which can impact the mechanical properties of the resulting materials [24].

3.3 COMPOSITES WITH NANOSCALE PLATELET REINFORCEMENT

In the review of nanoplatelet-reinforced composites, two types are examined: clay and graphite composites. Both clay and graphite materials possess a layered structure in their bulk form. To effectively utilize these components, the layers need to be separated and evenly distributed within the matrix phase [25,26].

In the case of clay particles, the interlayer gap is minimal in the conventional miscible state. However, when a polymer resin is introduced into the space between adjacent layers, known as the gallery, the spacing between the layers expands, leading to an intercalated state. The clay is considered to be exfoliated when the layers are completely separated from each other [26].

Similarly, in graphite composites, the layers of graphite are separated and distributed within the matrix material. The separation of graphite layers allows for enhanced properties such as improved mechanical strength and electrical conductivity [26].

The even distribution and separation of clay or graphite layers within the matrix are crucial for achieving optimal reinforcement effects in the composites. Techniques such as melt compounding, solution processing, or in situ synthesis are employed to achieve a uniform dispersion and exfoliation of the nanoplatelets within the composite matrix [27,28].

By utilizing clay and graphite nanoplatelets, researchers aim to develop composites with improved mechanical, thermal, and electrical properties, leading to their applications in various industries, including automotive, aerospace, electronics, and construction [27,28].

Clay materials, such as montmorillonite, saponite, and synthetic mica, are commonly used in the development of clay-based nanocomposites. Recent research has focused on advancing these clay-based nanocomposites and investigating their properties [27,28].

Polymer-based clay nanocomposites offer several benefits compared to traditional composites. They exhibit improved stiffness, strength, toughness, and thermal stability. Additionally, they demonstrate reduced gas permeability and coefficient of thermal expansion (CTE) [29,30].

One challenge in incorporating clay minerals into polymer matrices is the lack of affinity between the hydrophilic silicate surfaces of clay and the hydrophobic polymer matrix. This can result in agglomeration of the clay particles within the composite. To address this issue, surface modification of clay particles is carried out to enhance compatibility with the polymer matrix [30].

At Toyota Research Lab, pioneering research has shown that the addition of small amounts of montmorillonite clay material to polymer matrices can lead to significant improvements. These improvements include enhanced tensile strength, tensile modulus, and heat degradation temperature (HDT). Furthermore, clay-based nanocomposites exhibit reduced rates of water absorption and coefficient of thermal expansion in the flow direction. The incorporation of montmorillonite clay into polymer matrices has proven to be a promising approach for achieving enhanced mechanical and thermal properties in nanocomposites [30].

The surface modification of clay particles and their incorporation into polymer matrices open up possibilities for developing advanced materials with enhanced properties and performance. These clay-based nanocomposites have potential applications in various industries, including automotive, packaging, aerospace, and construction [31].

Exfoliated graphite or graphene sheets, similar to exfoliated clay, have a comparable thickness. However, they exhibit superior properties compared to clay platelets. Exfoliated graphite or graphene sheets have higher tensile modulus, tensile strength, and thermal conductivity, and lower electrical resistivity [31].

In polymer composites, achieving a certain weight content of the conductive phase results in a percolation threshold, beyond which the electrical resistance significantly decreases. Due to its low electrical resistance, exfoliated graphite enhances the conductivity of polymer composites when the percolation threshold is reached. Plasma treatment with O_2 is used to initiate radical polymerization on graphite nanoplatelets [31].

Exfoliated graphite nanoplatelet/thermoplastic composites have been observed to exhibit reduced percolation thresholds, with approximately 1% the threshold for exfoliated graphite. The addition of small amounts of graphite platelets significantly improves the electrical conductivity of polymeric materials, leading to various practical applications. These include electromagnetic interference (EMI) shielding, heat management in electronic and computer equipment, electrostatic paint for automobiles, and polymer cable sheathing [31,32].

The utilization of nanoclay/polypropylene composites in the automotive industry has been on the rise, with General Motors (GM) alone incorporating approximately 660,000 pounds of nanocomposites annually. This signifies the growing importance of these materials as functional parts in automobiles. By incorporating graphite and nanoclay into polymeric materials, the properties of the resulting composites are significantly improved. These advancements not only enhance the performance of composite materials but also expand their application potential across various industries. The ability to tailor the properties of composites through the incorporation of nanoclay and graphite opens up new possibilities for utilizing these materials in innovative and diverse ways [31,32].

3.4 RECENT DEVELOPMENTS IN NANOCOMPOSITE GRAPHENE-POLYMER

Graphene, a promising filler for nanocomposites, possesses unique multifunctional characteristics. However, the properties of graphene are influenced by various factors, including its 2D anisotropic nature, layer number, defects, size, functionalization, and matrix [33].

The last decade has witnessed significant advancements in both experimental and simulation studies focused on graphene nanocomposites. One of the critical factors influencing the performance of these composites is the interfacial configuration between graphene and the matrix material. The way graphene is connected to the matrix directly affects load transfer mechanisms, electrical conduction pathways, and thermal transport efficiency from graphene to the matrix [33,34].

Furthermore, the distribution of graphene within the matrix is a key aspect that governs the overall properties of nanocomposite materials. Achieving a uniform dispersion of graphene throughout the matrix is essential for realizing the full potential of its exceptional properties. The spatial arrangement and concentration of graphene nanosheets significantly impact the mechanical, electrical, and thermal behavior of the composite [33,34].

As a result, extensive research efforts have been directed toward understanding and optimizing the interfacial interactions and distribution of graphene in nanocomposite systems. This knowledge is vital for tailoring the properties of graphene-based nanocomposites to meet specific application requirements and unlock their full potential in various fields, including electronics, energy storage, and structural materials [34].

To fully leverage the 2D nature of graphene, considerable effort has been devoted to achieving a graphene network architecture and aligning graphene within nanocomposites. These advancements aim to enhance the overall performance and functionality of nanocomposite materials [34].

From a scientific and technical perspective, nanocomposites still pose certain challenges that require in-depth investigation. Some conflicts in nanocomposites need to be addressed, such as the trade off between strength and toughness in bulk nanocomposites, the balance between reflection and absorption in EMI nanocomposites, the interplay between thermal conductance and interfacial thermal resistance in thermal management nanocomposites, and the competition between conversion efficiency and thermal conductance in photothermal conversion nanocomposites [35].

While several graphene nanocomposites have been developed, most of them are currently limited to laboratory-scale applications. However, it is anticipated that the conflicts in mechanical properties, specifically the trade off between strength and toughness, can be resolved through the implementation of hierarchical and multi-scale architectures, as well as abundant interfacial interactions in graphene nano-composites [35].

Regarding the application of electrically conductive graphene nanocomposites for EMI shielding, it is preferable to prioritize electromagnetic absorption over reflection, as reflection is not environmentally friendly. Thus, the architectural design of graphene nanocomposites should focus on tailoring dielectric properties (such as gradient dielectric constants) to cover an ultra-broad frequency spectrum [36].

Furthermore, when enhancing the thermal properties of graphene nanocomposites, it is essential to also consider their mechanical properties, as both aspects need to meet the requirements of practical applications simultaneously [36]. Lastly, in the photothermal conversion of graphene nanocomposites, the speed of thermal conduction should be regulated in accordance with the specific applications [35,36]. Despite these challenges, there is strong confidence that these difficulties will be overcome, leading to the development of high-performance graphene nanocomposites and an accelerated commercialization process in the near future [36].

3.5   FIBER-REINFORCED NANOCOMPOSITES

Vapor-grown carbon nanofibers (CNF) have emerged as versatile reinforcements for various polymer matrices. They have been successfully incorporated into a wide range of polymers, including polycarbonate, nylon, poly(ethylene terephthalate), polypropylene, poly(phenylene sulfide), poly(ether sulfone), acrylonitrile-butadiene-styrene (ABS), and epoxy [36,37].

Carbon nanofibers exhibit diverse morphologies, which contribute to their unique properties. These morphologies can range from disordered bamboo-like structures to highly graphitized "cup-stacked" structures, where conical shells are nested within one another. The arrangement of carbon atoms within the nanofibers determines their structural characteristics and influences their mechanical, electrical, and thermal properties [36,37].

In terms of diameter, carbon nanofibers typically range from 50 to 200 nm, offering a nanoscale reinforcement option for polymer composites. The specific diameter and morphology of nanofibers can be tailored to achieve desired composite properties and optimize the interface between the nanofibers and the polymer matrix [36,37].

The incorporation of carbon nanofibers into polymer matrices has been shown to enhance the mechanical strength, stiffness, electrical conductivity, and thermal stability of the resulting nanocomposites. These improvements arise from the high aspect ratio, excellent tensile properties, and superior interfacial bonding between nanofibers and the polymer matrix [37].

Overall, vapor-grown carbon nanofibers serve as a promising reinforcement material in polymer composites, offering a wide range of possibilities for enhancing the performance of various polymer-based applications [37].

The study conducted by Wei and Srivastava using continuum elastic theory and molecular dynamics simulations shed light on the relationship between the mechanical properties of carbon nanofibers and their morphologies, specifically focusing on the axial Young's modulus [37]. The researchers discovered that the axial Young's modulus of carbon nanofibers is strongly influenced by the shell tilt angle, which refers to the angle at which the graphene layers in the nanofiber are tilted with respect to the fiber axis. They observed that nanofibers with smaller axial tilt angles displayed significantly higher Young's moduli compared to those with larger tilt angles [37].

The sensitivity of the Young's modulus to the shell tilt angle suggests that the arrangement and orientation of graphene layers within the nanofiber play a crucial role in determining its mechanical properties. Nanofibers with smaller tilt angles likely have a more ordered and aligned structure, resulting in enhanced stiffness and a higher Young's modulus [36].

These findings highlight the importance of understanding the structural characteristics of carbon nanofibers and their impact on mechanical properties. By controlling the morphologies and orientations of the graphene layers in carbon nanofibers, it may be possible to tailor their mechanical properties to suit specific applications, enabling the design of advanced materials with desired performance characteristics [37].

Due to the diverse morphology of carbon nanofibers and their associated properties, experimental results on the processing and characterization of nanofiber composites demonstrate a wide range of scatter. This variability highlights the importance of considering the specific morphological characteristics of carbon nanofibers when studying and utilizing them as reinforcements in composite materials [38].

In the production of carbon nanofibers, longer gas-phase feedstock residence periods have been found to result in the production of nanofibers that are less graphitic. However, these nanofibers exhibit better adhesion to the polypropylene matrix. As a result, composites incorporating these nanofibers demonstrate improved tensile strength and Young's modulus [38].

The oxidation of carbon nanofibers has been identified as a means to enhance the tensile strength and adherence of composites to the matrix. However, it is important to note that prolonged oxidation can have a negative impact on the properties of both the fibers and the resulting composites [38].

The production, characterization, and modeling of carbon nanofiber composites face similar challenges to those encountered in nanotube-reinforced composites. These challenges include achieving the proper dispersion of the nanofibers within the matrix and ensuring strong adhesion between the fibers and the matrix. Overcoming these challenges is crucial for optimizing the performance and properties of carbon nanofiber composites [37,38].

3.6   COMPOSITES REINFORCED WITH CARBON NANOTUBES

The physical form of a carbon nanotube is determined by the direction and magnitude of its chiral vector, which is derived from "wrapping up" a graphene sheet. There are two limiting designs of CNTs known as armchair nanotubes and zigzag nanotubes. The chiral vector defines the structure and properties of the nanotube [4,5].

Indeed, carbon nanotubes possess exceptional thermal and electrical characteristics, high specific stiffness and strength, as well as high aspect ratios, which have sparked significant interest in their application as reinforcements in composite materials [4–6].

The incorporation of CNTs into composite matrices offers the opportunity to enhance the various properties of the resulting nanotube-reinforced composites. One notable improvement is observed in mechanical strength, where the high tensile strength of CNTs can significantly enhance the overall strength of the composite material. The exceptional stiffness of CNTs also contributes to improved specific stiffness, allowing for the creation of lightweight yet robust composite structures [7,8].

CNTs also exhibit outstanding thermal conductivity, which can be advantageous in applications requiring efficient heat dissipation. By dispersing CNTs within a composite matrix, the thermal conductivity of the material can be greatly enhanced, enabling better thermal management and heat transfer [7–9].

Furthermore, the excellent electrical conductivity of CNTs allows for the development of composites with enhanced electrical properties. The presence of CNTs in the composite matrix enables the conduction of an electrical charge, opening up possibilities for applications in electronic devices, sensors, and conductive materials [8,9].

The high aspect ratio of CNTs, combined with their unique structural characteristics, provides an effective load transfer pathway within the composite material. This results in improved interfacial bonding between the CNTs and the matrix, leading to enhanced overall performance of the composite in terms of mechanical, thermal, and electrical properties [11].

Overall, the exceptional properties of CNTs make them highly desirable as reinforcing agents in nanotube-reinforced composites. These composites have the potential to revolutionize various industries by offering improved mechanical strength, thermal conductivity, electrical conductivity, and other performance parameters, paving the way for advanced structural and functional applications [12].

CNT-reinforced composites have demonstrated significant potential in a variety of industries, including aerospace, automotive, electronics, and energy. The combination of the remarkable properties of CNTs and their ability to enhance the performance of the matrix material has led to extensive research and exploration of their applications in different fields [14].

3.7 NANOMATERIALS FOR BONE TISSUE REGENERATION

Nanocomposite biomaterials are an emerging field of study that combines bioactive nanosized fillers into a biopolymeric and biodegradable matrix structure. Nanocomposites can be roughly classified into two categories: those based on natural polymers and those based on synthetic polymers [8].

Nanocomposites based on natural biopolymers such as chitosan (CS), collagen (Col), cellulose, silk fibroin (SF), alginate, and fucoidan are produced by incorporating nanoparticles and/or nanofibers into the matrix. Polycaprolactone (PCL), poly(lactic-co-glycolic acid) (PLGA), polyethylene glycol (PEG), and poly(lactic acid) (PLA) are examples of synthetic polymers used in nanocomposites [7].

Nano-hydroxyapatite (nHA), nano zirconia (nZr), nano silica (nSi), nano silver nanoparticles (AgNPs), nano titanium dioxide (nTiO$_2$), and nano graphene oxide (GO) are just a few of the nanofillers extensively employed with natural and/or synthetic polymer matrices in bone tissue regeneration research. The nanofillers boost the mechanical strength and stability of the nanocomposites and promote cell adhesion, proliferation, and differentiation [8–10].

Bone regeneration, cell proliferation, and adhesion can all be aided by biodegradable polymer nanocomposites. Combining physical, chemical, and biological qualities that are similar to the bone's natural extracellular matrix (ECM) is made possible by incorporating nanofillers into the polymer matrix. Growth elements and nutrients are transported more efficiently and new bone tissue is formed more quickly because of this hybrid structure's optimum porosity, mechanical strength, and flexibility [9].

Many factors and techniques for synthesis, analysis, and measurement go into making these nanocomposites. Improving the biomaterial as a whole requires fine-tuning the material preparation and mixing processes. When nanofillers are added to a hydrogel-based polymer matrix, it can be used as an injectable scaffold. This gives it an edge over 3D scaffolds and makes the process of implanting a scaffold to fix craniofacial deformities less intrusive [9].

When it comes to integrating with the host bone, a fast breakdown rate and the smooth removal of degradation products from the scaffold are both dependent on the cross-linking density of the polymer matrix during the biodegradation phase. Synthetic polymers permit the improvement of material qualities via physical, chemical, and topographical means, while natural polymers provide advantages in biocompatibility and degradation. Acid buildup in the surrounding tissue is one problem that might result from the degradation of the byproducts of synthetic polymer–based scaffolds. Thus, it is essential to enhance polymer blends with nanofillers to stimulate healthy bone tissue formation [17–19].

Increased cytocompatibility, bioactivity, and bone tissue development make nanocomposite biomaterials an attractive area for future research in the field of bone tissue regeneration [19].

3.8   BIOMATERIALS AND DENTAL IMPLANTS

Creating dental implants with the desired characteristics remains a challenging task, despite the promising research conducted in nanotechnology for implant surface engineering. It is crucial to carefully evaluate potential risks to human health associated with nanoparticles, such as nanotoxicity, and consider various other factors including pH changes in the surrounding tissue and the metabolic activities of the body. Balancing the properties of osseointegration, biodegradability, and biocompatibility, and avoiding negative impacts on biological structures are a complex endeavor [16–18].

In this regard, materials derived from natural resources such as hydroxyapatite, collagen, and gelatin should be given more attention, as they can be modified to achieve desired surface properties. Moreover, the utilization of computational models and machine learning technologies can significantly enhance biomaterial design by predicting interactions with biological structures and assessing the extent of healing. Therefore, focused and critical studies are necessary to develop the ideal medical implant [17–19].

Current research findings in this field are highly promising and have the potential to improve the quality of life for individuals through advancements in medical implant technology [18,19].

3.9 NANOCOMPOSITES' CRITICAL CHALLENGES

The major challenges in the study of nanocomposites can be classified into three categories: structural characterization, interfacial bonding, and processing [31].

3.9.1 STRUCTURAL CHARACTERIZATION

Theoretical modeling and experimental characterization of nanoscale reinforcing materials, such as nanotubes, present significant challenges. The behavior of nanoscale materials differs from bulk materials, and their properties can be influenced by factors such as size, morphology, defects, and functionalization. Understanding and accurately characterizing these properties are crucial for predicting the performance of nanocomposites [34].

3.9.2 INTERFACIAL BONDING

The interfacial bonding between the reinforcing nanoparticles and the matrix material is critical for the mechanical properties of nanocomposites. Achieving a strong and efficient interface between the two phases is essential to effectively transfer loads and enhance mechanical performance. However, the understanding of interfacial bonding mechanisms is still limited, both analytically and experimentally. Developing techniques to improve interfacial bonding and maximize reinforcement efficiency is an ongoing challenge [35].

3.9.3 PROCESSING

The processing of nanocomposites poses several challenges. Achieving a uniform dispersion of nanoparticles within the matrix material is crucial for optimizing the properties of the resulting composite. However, nanoparticles tend to agglomerate, leading to poor dispersion and uneven mechanical properties. Additionally, the processing techniques need to be compatible with the nanomaterials and should not cause degradation or damage to the nanoparticles. Developing effective processing methods that can overcome these challenges and ensure uniform dispersion is vital for the successful fabrication of nanocomposites [36].

Overall, addressing the challenges related to structural characterization, interfacial bonding, and processing is essential for advancing the field of nanocomposites and realizing their full potential in various applications. Continued research and development efforts are required to overcome these obstacles and enable the wide-spread use of nanocomposites in diverse industries [36].

3.9.4 Dispersion

In the processing of nanocomposites, one of the primary challenges is achieving a uniform dispersion of nanoparticles and nanotubes to prevent their agglomeration, which is primarily caused by van der Waals bonding. Agglomeration can signifi-cantly affect the mechanical and physical properties of the resulting nanocomposite [35,36].

In addition to preventing nanoparticle agglomeration, the exfoliation of clays and graphitic layers is crucial in nanocomposite processing. Clays and layered materi-als such as graphene have a tendency to form stacked structures, and it is essential to separate these layers to fully utilize their properties in the composite material [35,36].

When dealing with carbon nanotubes, specific challenges arise. Single-walled carbon nanotubes (SWCNTs) tend to form rope-like clusters, while multi-walled carbon nanotubes (MWCNTs) produced through chemical vapor deposition often tangle together like spaghetti strands. To enable proper alignment and dis-persion, nanotubes need to be effectively separated in a solvent or matrix mate-rial [35,36].

Various techniques are employed to address these challenges and achieve the uni-form dispersion of nanoparticles and nanotubes. These include sonication, surfac-tant-assisted dispersion, functionalization of nanoparticles, and the use of specific solvents or polymer matrices that are compatible with the nanomaterials. These tech-niques aim to overcome the attractive forces between nanoparticles or nanotubes and disperse them evenly within the matrix material [35,36].

The successful dispersion and separation of nanoparticles and nanotubes are crucial for achieving the desired properties and performance of nanocomposites. Proper processing techniques are vital to ensure that the nanoparticles are uniformly distributed and effectively interact with the matrix material, leading to improved mechanical, electrical, and thermal properties [35–37].

Continued research and development efforts are focused on optimizing dispersion techniques and developing novel methods to address the challenges associated with agglomeration and exfoliation. These advancements in nanocomposite processing will enable the production of high-performance materials with enhanced properties for various applications [37].

3.9.5 Alignment

The lack of control over the orientation of nanotubes in a polymeric matrix mate-rial poses a significant challenge in achieving efficient nanotube reinforcement, in terms of both structural and functional performance. Due to their small sizes, align-ing nanotubes in a manner similar to traditional short fiber composites is extremely challenging [37,38].

To optimize the reinforcement capabilities of nanotubes, it is essential to achieve controlled alignment within the composite material. However, the inherent characteristics of nanotubes make it difficult to effectively manipulate their orientation. Currently, there is limited control over the alignment of nanotubes, which hinders their full potential in enhancing the properties of nanocomposites [37,38].

Efforts are being made to develop innovative processing techniques and methods to overcome these challenges. Researchers are exploring various approaches such as electric fields, magnetic fields, shear forces, and templates to guide the alignment of nanotubes within the polymer matrix. These techniques aim to improve the dispersion and alignment of nanotubes, enabling a more efficient transfer of mechanical, electrical, and thermal properties [37,38].

By achieving better control over the orientation of nanotubes in the composite, it becomes possible to fully exploit their unique characteristics. Aligned nanotubes can enhance the mechanical strength, electrical conductivity, and thermal conductivity of the resulting nanocomposite, making them suitable for a wide range of applications [37,38].

Continued research and advancements in nanotube alignment techniques will lead to improved control over their orientation and enable the development of high-performance nanocomposites with tailored properties [37,38].

3.9.6 VOLUME AND RATE

Efficiency in manufacturing plays a critical role in the development and commercialization of nanocomposites as viable products. The lessons learned from the fabrication of traditional fiber composites have highlighted the importance of establishing a strong scientific foundation for manufacturing processes [32].

In order to make nanocomposites commercially viable, it is essential to achieve high volume and high rate fabrication capabilities. This involves developing efficient and scalable manufacturing techniques that can produce nanocomposites in large quantities and at a rapid pace [32].

Efficiency in manufacturing encompasses several key aspects. Firstly, it involves optimizing the dispersion and alignment of nanoparticles or nanotubes within the matrix material. Achieving a uniform dispersion and controlled alignment is crucial for maximizing the reinforcement potential and properties of nanocomposite [31,32].

Secondly, efficient manufacturing involves selecting suitable processing methods that can handle the unique properties and challenges associated with nanomaterials. This includes considerations such as compatibility between the matrix material and nanoparticles, efficient mixing and blending techniques, and precise control over processing parameters [31,32].

Additionally, process optimization and automation can significantly enhance manufacturing efficiency. By streamlining production processes, minimizing waste, and implementing quality control measures, the fabrication of nanocomposites can be more cost-effective and consistent [31,32].

Furthermore, the development of reliable and standardized testing and characterization methods specific to nanocomposites is essential. This allows for the accurate assessment of their properties and performance, ensuring quality control and facilitating the commercialization of nanocomposite products [30–32].

Overall, efficiency in manufacturing is a crucial factor in the future development of nanocomposites. By establishing a robust science base and leveraging lessons learned from traditional fiber composites, researchers and industry professionals can advance manufacturing techniques to meet the demands of high volume and high rate fabrication. This, in turn, will contribute to the widespread adoption and commercial success of nanocomposites in various industries and applications [31,32].

3.9.7   Cost-Effectiveness

The cost of nanocomposites is influenced by several factors, including the cost of the nano reinforcement material, such as nanotubes. Nanotubes, one of the commonly used nano reinforcements, contribute to the overall cost of nanocomposites. However, as the applications for nanocomposites continue to expand, the cost of nanotubes and their composites is expected to decrease significantly. This is primarily driven by economies of scale and advancements in manufacturing techniques that enable high volume and high rate production [27,28].

The properties of nanostructured composites, including nanocomposites, are strongly influenced by their size and structure. Understanding the properties and interactions across different length scales is crucial to harness the exceptional qualities observed at the nanoscale and effectively utilize them at the macroscale. Developing a fundamental comprehension of structure–property relationships allows for the design and engineering of multifunctional nanoscale materials suitable for various applications, ranging from structural and functional materials to biomaterials [28,29].

Scaling up manufacturing techniques is essential for the large-scale application of nanocomposites. The ability to produce nanocomposites in high volumes and at high rates is crucial for their commercial viability. Researchers and industry professionals are continuously working on developing efficient and scalable manufacturing processes to meet the growing demand for nanocomposites [28,29].

In addition to technical considerations, the broad societal effects of nanotechnology, including nanocomposites, need to be addressed. As with any emerging technology, it is important to assess and mitigate potential environmental, health, and safety risks associated with the production, use, and disposal of nanocomposites. Responsible development and regulation of nanocomposites are necessary to ensure their safe and sustainable integration into various industries and applications [29–31].

In summary, the cost of nanocomposites is influenced by factors such as the cost of nano reinforcements, particularly nanotubes, as well as the efficiency of high volume and high rate production. Understanding the structure–property relationships across different length scales is crucial for optimizing the properties of nanocomposites. Scaling up manufacturing techniques and addressing the broader societal impacts of nanocomposites are also key considerations for their successful implementation [29–31].

REFERENCES

1. R. Florencio-Silva, G.R.D. Sasso, E. Sasso-Cerri, M.J. Simoes, P.S. Cerri, Biology of bone tissue: Structure, function, and factors that influence bone cells, *BioMed Res. Int.* (2015). https://doi.org/10.1155/2015/421746.

2. A.R. Amini, C.T. Laurencin, S.P. Nukavarapu, Bone tissue engineering: Recent advances and challenges, *Crit. Rev. Biomed. Eng.* 40(5) (2012) 363–408.

3. R.J. O'Keefe, J. Mao, Bone tissue engineering and regeneration: From discovery to the clinic—An overview, *Tissue Eng. B* 17(6) (2011) 389–392.

4. C.L.M. Bao, E.Y. Teo, M.S. Chong, Y. Liu, M. Choolani, J.K. Chan, Advances in bone tissue engineering, In Regenerative Medicine and Tissue Engineering, *InTech* (2013) 600–614. http://doi.org/10.5772/55916.

5. B. Baroli, From natural bone grafts to tissue engineering therapeutics: Brainstorming on pharmaceutical formulative requirements and challenges, *J. Pharm. Sci.* 98(4) (2009) 1317–1375.

6. J. Parvizi, *High Yield Orthopaedics E-book*, Elsevier Health Sciences, 2010. ISBN no: 978-1416002369, Saunders/Elsevier, Philadelphia.

7. R. Mishra, T. Bishop, I.L. Valerio, J.P. Fisher, D. Dean, The potential impact of bone tissue engineering in the clinic, *Regen. Med.* 11(6) (2016) 571–587.

8. R. Dimitriou, E. Jones, D. McGonagle, P.V. Giannoudis, Bone regeneration: Current concepts and future directions, *BMC Med.* 9(1) (2011) 66.

9. M.J. Yaszemski, R.G. Payne, W.C. Hayes, R. Langer, A.G. Mikos, Evolution of bone transplantation: Molecular, cellular and tissue strategies to engineer human bone, *Biomaterials* 17(2) (1996) 175.

10. A. Bharadwaza, A.C. Jayasuriya, Recent trends in the application of widely used natural and synthetic polymer nanocomposites in bone tissue regeneration, *Mater. Sci. Eng. C Mater. Biol. Appl.* 110 (2020) 110698.

11. S. Gautam, D. Bhatnagar, D. Bansal et al., Recent advancements in nanomaterials for biomedical implants, *Biomed. Eng. Adv.* 3 (2022) 100029.

12. A. Bandyopadhyay et al., Improving biocompatibility for next generation of metallic implants, *Prog. Mater. Sci.* 133 (2023) 101053.

13. K.K. Amirtharaj Mosas, A.R. Chandrasekar, A. Dasan, A. Pakseresht, D. Galusek, Recent advancements in materials and coatings for biomedical implants, *Gels* 8(5) (2022) 323.

14. Dr. X. Sun, Dr. L. Liang, Dr. Y. Cheng, Prof. Y. Li, Recent progress in graphene/polymer nanocomposites, *Adv. Mater.* 33 (2020) 2001105.

15. I.A. Kinloch, J. Suhr, J. Lou, R.J. Young, P.M. Ajayan, Composites with carbon nanotubes and graphene: An outlook, *Science* 362(6414) (2018) 547.

16. S.-S. Yao, F.-L. Jin, K.Y. Rhee, D. Hui, S.-J. Park, Recent advances in carbon-fiber-reinforced thermoplastic composites: A review, *Composites Part B* 142 (2018) 241.

17. Z. Yin, Q. Zheng, CuO/polypyrrole core–shell nanocomposites as anode materials for lithium-ion batteries, *Adv. Energy Mater.* 2(2) (2012) 179.

18. X. Yu, H. Cheng, M. Zhang, Y. Zhao, L. Qu, G. Shi, Graphene-based smart materials, *Nat. Rev. Mater.* 2(9) (2017) 17046.

19. A.K. Naskar, J.K. Keum, R.G. Boeman, Effects of nano reinforcing/matrix interaction on chemical, thermal and mechanical properties of epoxy nanocomposites, *Nat. Nanotechnol.* 11(12) (2016) 1026.

20. L.V. Kayser, D.J. Lipomi, Stretchable Conductive Polymers and Composites Based on PEDOT and PEDOT:PSS, *Adv. Mater.* 31(10) (2019) 1806133.

21. E. Lizundia, I. Serna, E. Axpe, J.L. Vilas, Free-volume effects on the thermomechanical performance of epoxy–SiO2 nanocomposite, *J. Appl. Polym. Sci.* 134(34) (2017) 45216.

22. E. Alishahi, S. Shadlou, S. Doagou-R, M.R. Ayatollahi, Effects of carbon nanorein-forcements of different shapes on the mechanical properties of epoxy-based nanocom-posites, *Macromol. Mater. Eng.* 298(6) (2013) 670.

23. S. Sharma, N.C. Kothiyal, Synergistic effect of zero-dimensional spherical carbon nanoparticles and one-dimensional carbon nanotubes on properties of cement-based ceramic matrix: Microstructural perspectives and crystallization, *Compos. Interfaces* 22(9) (2015) 899.

24. R. Adams, A review of the stainless steel surface, A review of the stainless steel sur-face, *J. Vac. Sci. Technol. A* 1(1) (1983) 12.

25. H.A. Alaraby, M. Iswalhia, T. Ahmed, A study of mechanical properties of titanium alloy Ti-4A1-4V used as dental implant material. *Int. J, Sci. Rep.* 3 (2017) 288–291.

26. D. Amalraju, A. Dawood, Mechanical strength evaluation analysis of stainless steel and titanium locking plate for femur bone fracture, *Int. J. Eng. Sci. Technol.* 2(3) (2012) 381.

27. A. Anders, Metal plasma immersion ion implantation and deposition: A review, *Surf. Coat. Technol.* 93(2–3) (1997) 158–167.

28. D. Arcos, M. Vallet-Regí, Substituted hydroxyapatite coatings of bone implants, *J. Mater. Chem. B* 8(9) (2020) 1781–1800.

29. T. Aviles, S.-M. Hsu, A. Clark, F. Ren, C. Fares, P.H. Carey, J.F. Esquivel-Upshaw, Hydroxyapatite formation on coated titanium implants submerged in simulated body fluid, *Materials (Basel)* 13(24) (2020) 5593.

30. G.C. Babis, A.F. Mavrogenis, Cobalt–chrome porous-coated implant-bone interface in total joint arthroplasty. In *Bone-Implant Interface in Orthopedic Surgery*, London: Springer, 2014, 55–65.

31. M. Balazic, J. Kopac, M.J. Jackson, W. Ahmed, Review: Titanium and titanium alloy applications in medicine, *Int. J. Nano Biomater.* 1(1) (2007) 3–34.

32. V.K. Balla, S. Bodhak, S. Bose, A. Bandyopadhyay, Porous tantalum structures for bone implants: Fabrication, mechanical and in vitro biological properties, *Acta Biomater.* 6(8) (2010) 3349–3359.

33. V.K. Balla, S. Bose, N.M. Davies, A. Bandyopadhyay, Tantalum–a bioactive metal for implants, *JOM* 62(7) (2010) 61–64.

34. G. Tang, Z. Liu, Y. Liu, J. Yu, X. Wang, Z. Tan, X. Ye, Recent trends in the development of bone regenerative biomaterials, *Front. Cell Dev. Biol.* 9 (2021) 1001.

35. A.A. Campbell, Bioceramics for implant coatings, *Mater. Today* 6(11) (2003) 26–30.

36. A. Goharian, Fundamentals in loosening and osseointegration of orthopedic implants, *Osseointegration Orthop. Implant* 1 (2019) 1–26.

37. T. Rodriguez-Gabella, P. Voisine, R. Puri, P. Pibarot, J. Rodés-Cabau, Aortic biopros-thetic valve durability: Incidence, mechanisms, predictors, and management of surgical and transcatheter valve degeneration, *J. Am. Coll. Cardiol.* 70(8) (2017) 1013–1028.

38. K. Prasad, O. Bazaka, M. Chua, M. Rochford, L. Fedrick, J. Spoor, R. Symes, M. Tieppo, C. Collins, A. Cao et al., Metallic biomaterials: Current challenges and oppor-tunities, *Materials (Basel)* 10(8) (2017) 884.

4 Medical Imaging for Patient-Specific Implants

Radha Sarawagi Gupta, Babu Lal,
Abhinav Chander Bhagat, and Ragavi Alagarsamy

4.1 INTRODUCTION

Advances in modern healthcare and technology have ushered in a revolution in the field of patient-specific healthcare solutions. A patient-specific implant (PSI) is one such novel product of technological progress. These are custom-made implants (CMIs) or custom-made prosthesis (CMP) manufactured to the individual patient's anatomic specifications to address functional or cosmetic limitations following an accident or disease. They provide a customized fit with greater accuracy along with the benefits of a shorter surgical time, shorter rehabilitation period, and reduced overall financial burden.[1]

The journey of PSI fabrication starts with imaging, followed by designing and manufacturing. Each step is crucial and should be executed following certain preset principles for optimal outcomes and the long-term success of PSIs. Imaging such as a computed tomography (CT) scan or magnetic resonance imaging (MRI) with standard protocol should be considered for any PSI planning. Designing is a didactic process between the biomedical engineer and the treating surgeon, requiring multiple cyclic iterations. Special methods such as finite element analysis (FEA) and topology optimization (TO) are add-on steps for design optimization. After the treating surgeon's final approval of the design, fabrication can be carried out either by computer numerical control (CNC) machining or 3D printing, or a hybrid process depending on the type, complexity, and availability of manufacturing services. This chapter focuses on the main steps of PSIs fabrication, from imaging protocol to final implantable PSIs.

4.2 PSI SCOPE AND EVIDENCE

Using autologous means to restore the bony or soft tissue may not always be feasible due to donor-site morbidity, limited graft size, patient's medical condition, and willingness, among others. Furthermore, carving an autologous graft to the desired form is challenging; the fit and contour may be imprecise, leading to potential risks of resorption and healing.[2] As an alternative solution, the PSI or patient-specific prosthesis (PSP) may be utilized in such patients where off-the-shelf implants have

DOI: 10.1201/9781003375098-4

yielded unacceptable results.[3-5] PSIs can be combined with grafts or flaps for augmented benefits.

The scope of PSIs is not limited to bony and reconstructive surgeries in specialties such as orthopedics and maxillofacial surgery, but also facilitates augmentation of the soft tissues. In addition, PSIs enable partial or total organ replacement in the field of prosthetics and orthotics. Literature evidence suggests that PSIs have gained wide acceptance in various medical fields, such as oral and maxillofacial surgery, orthopedics, dentistry, neurosurgery, and orthotic and prosthetics, as demonstrated in Figure 4.1. PSIs can be a good alternative when the bony or soft tissue defect is secondary to congenital or acquired conditions such as post-trauma, burns, pathology, cancer, or complications resulting from surgical or therapeutic interventions. The following section presents an overview of the evidence-based scope of PSIs or PSPs in diverse medical fields and anatomical regions of the human body.

4.2.1 NEUROSURGERY

Various traumatic and pathological processes can cause cranial bone defects. These include post-traumatic fractures and various benign and malignant neoplasms,

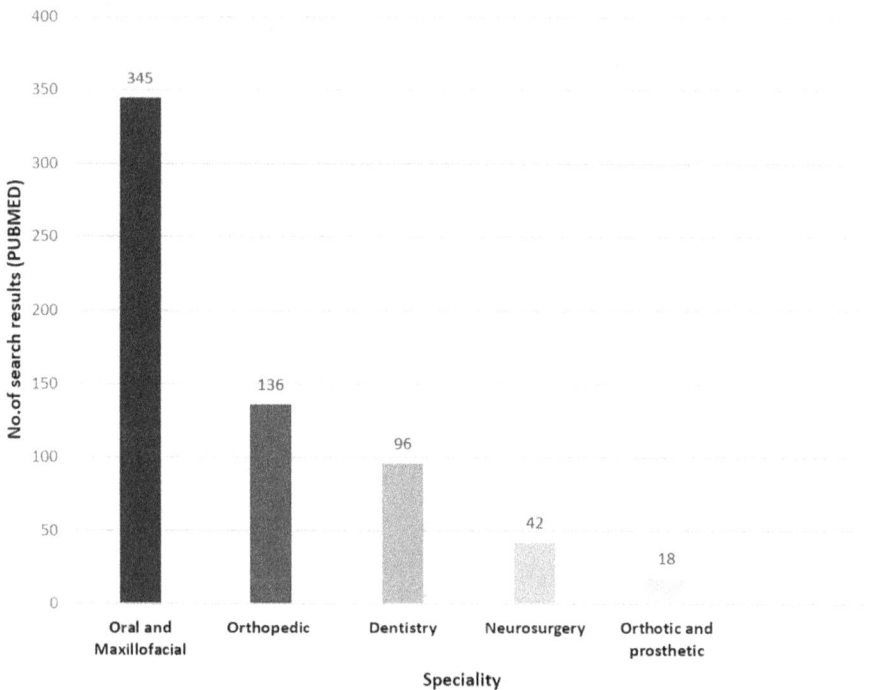

FIGURE 4.1   Graph illustrating publications related to PSIs among various medical specialties. Specialty wise overview of PSI-related search results (Based on PubMed database search since inception to April 2023).

including Langerhans cell histiocytosis, hemangioma, fibrous dysplasia, meningioma, primary bone sarcoma, and metastasis. Bone infections (primary or secondary) can also result in calvarial defects. Regardless of the underlying cause, these defects result in functional issues if left unaddressed and a cosmetic component that may negatively impact the patient's self-image. Conventional implants used in such cases include autologous bone grafts, bone cement, and malleable titanium meshes, which help restore mechanical functions; however, these suffer from the drawback of suboptimal cosmesis and only partial restoration of the anatomy. In these situations, PSIs play an essential role in restoring mechanical protection and establishing an excellent cosmetic appearance.[6]

4.2.2 MAXILLOFACIAL SURGERY

Maxillofacial defects can be both congenital and acquired in nature. The complex anatomy of this region and its effect on the patient's appearance render these a challenge.[7] The surgeon has to take care of the functional, mechanical, aesthetic, and psychological components. Many surgeons still prefer autogenous grafts; nevertheless, the resorption rate is unpredictable and is accompanied by donor-site morbidity. Traditional premade implants require many adjustments and offer suboptimal results.[8,9] PSIs, on the contrary, have been found to have higher geometrical accuracy, optimal biomechanical properties, better stability, shortened recovery time, and enhanced facial contour preservation with reduced postoperative complications.[2,10,11]

Various types of PSIs commonly used in the maxillofacial region include prostheses for facial defects and CMIs for orbital reconstruction[12,13] and mandibular and maxillary defects.[14–17] Augmented prostheses for the anterior mandible, angle of mandible, and zygomatic region are also used.[18] In recent years, the use of PSIs in orthognathic surgery has eliminated cumbersome lab procedures and provided more accurate results for maxillary repositioning.[19,20] The PSI has been validated in the temporomandibular joint region, including both mandibular and fossa components. Evidence suggests that the PSI has demonstrated acceptable long-term outcomes and required fewer revision surgeries than standard stock implants.[5,13,21–25] Recently, 3D-printed biodegradable polymers with good strength have been introduced to the market; these are a boon for customized implants in the management of fractures.[19]

4.2.3 ORTHOPEDIC SURGERY

PSIs have found useful application in managing complex bone defects in numerous orthopedic disciplines, including trauma, arthroplasty, tumor surgery, deformity correction, and osteochondral defect implant design.[26,27] The resection of a bone tumor results in a defect that requires reconstruction to restore stability and function. For instance, in spine cancer surgery and tumor removal, patient-specific vertebral prostheses have shown the best outcome with the preservation of structural integrity and spinal stability.[28–30] Custom implants are useful for reconstruction in clavicular, scapular, and pelvic bone tumors, as they have a unique bone geometry; an off-the-shelf prosthesis is not available.[31,32] In knee joint surgeries, PSIs have revealed better daily function and higher patient satisfaction compared to standard prostheses. In addition, they resulted in better kinematics like native knee joints and better contact stress compared to stock prostheses. Using CT images, the implant is bespoke designed to match the patient's knee shape and contour, preserving the bone.

Customized tools such as a patient-specific drill guide template help protect vital structures such as the spinal cord, dural sac, and neurovascular bundles during bone cutting and tumor removal surgery. They enhance accuracy and efficiency, reducing overall surgery time and safety in lowering intraoperative radiation.[33] In addition, customized instruments that facilitate accurate needle positioning for cryotherapy and help protect adjacent tissue from any thermal injury have been used.[34]

4.2.4 DENTISTRY

Computer-aided designing (CAD) and computer-aided manufacturing (CAM) technology have facilitated the application of PSIs in different specialties of dentistry. Customized dentures and associated PSI components have revolutionized the field of prosthodontics. A zygoma PSI is considered a better alternative in cases that lack bone constructs for denture support, and grafting increases the morbidity or is not feasible. A PSI is also good in large maxillary defects secondary to oral cancer, oral pathology, or post-mucormycotic cases.[35] 3D-printed crowns, bridges, and space maintainers for the pediatric population can be executed seamlessly.

4.2.5 ORTHOTICS AND PROSTHETICS

CAD/CAM technology has revolutionized the orthotic and prosthetic fields by enabling highly customized and personalized medical devices. Prosthetic needs in children are complex due to their changing anatomy and small size. CAD/CAM and 3D scanning of the human parts have facilitated orthotic device manufacture with a decreased lead time and cost. Customized prostheses for the rehabilitation of the foot, limbs, and arm, and braces for the hand, chest, and neck can be manufactured by 3D printing.[36–38]

4.3 IMAGING FOR PSI

Imaging is the first step for PSI generation. Various types of imaging such as a CT scan, an MRI, or a cone-beam CT (CBCT) scan, and surface 3D scanning are used to acquire the three-dimensional data of the region of interest. The selection of the modality of imaging depends on the individualized requirement of the case.

4.3.1 ROLE OF IMAGING

One of the major factors that have enabled the success of PSIs in modern healthcare has been the advancement in contemporary medical imaging technology. Imaging is crucial in various phases of implant surgery, from diagnosis and pathology evaluation to postoperative evaluation. The objective of any imaging and the choice of modality depend on multiple factors such as the body part to be imaged, the amount and type of information required, and the phase of the treatment procedure, i.e., preoperative, intraoperative, or postoperative evaluation.[39]

4.3.2 Imaging Modalities

Various imaging modalities such as x-ray, CT/CBCT, and MRI play an important role during pre- and post-implant imaging, of which thin-section high-resolution CT remains the workhorse.

4.3.2.1 Plain Radiography

Plain radiography is a readily available modality but gives limited details of the bony anatomy and pathology due to the superimposition of anatomical structures. It is commonly used in dental implant placements. Preoperative planning can be performed with the templating system, such as in hip arthroplasty. However, due to the 2D nature of this modality, accurate patient positioning and measurements are difficult to achieve. Also, the magnification factor of a digital radiograph is imprecise.[40]

4.3.2.2 Computed Tomography

Computed tomography is one of the most commonly used modalities for the preoperative planning of PSIs. It is a cross-sectional imaging modality that works on the principle of x-ray beam attenuation by the body tissues. X-ray projections through the patient from many different angles are received in a detector. The information is digitally combined to produce a cross-sectional body part image. With newer CT scanners featuring spiral technology and multi-slice detectors, a volume of data is acquired and can be reconstructed in any plane. A CT scan has several advantages over 2D radiographs. The examination speed is high, accurate volume coverage is possible, and it gives high-resolution images. CT is the imaging modality of choice when bones are the region of interest. A CT can acquire thin-slice axial images (slice thickness <1 mm) with isotropic voxels. It is readily available in most places and is also cost-effective. Precise measurement is possible, and data can be easily processed by software.

Radiation exposure is one of the limitations; however, with newer scanners featuring various dose reduction techniques such as tube current modulation and iterative reconstruction, there is a significant radiation dose reduction to the patient.[41] However, another drawback is CT's relatively poor soft tissue resolution compared to MRI.

CBCT is a variation of conventional CT scans used in dental and maxillofacial imaging. The x-ray tube and detector array rotate around the patient's head and use a cone-shaped x-ray beam to acquire images. A major advantage is a lower radiation dose as compared to conventional CT scans. It is the preferred modality in dental implant imaging.[42]

4.3.2.3 Magnetic Resonance Imaging

MRI works on the principle of a nuclear magnetic resonance (NMR) signal. Magnetic field gradients are used to spatially encode the NMR signal and acquire raw data. Various mathematical calculations, such as Fourier transformation, transform this raw data (k-space) into an image.

MRI has excellent soft tissue contrast and gives detailed information on the bone and soft tissue interface. Hence, MRI is ideally suited for presurgical planning of the extent of tumor resection in patients with bone tumors. MRI is also a beneficial modality for the evaluation of implant-related complications. It is superior to CT in defining soft tissue anatomies, such as articular cartilage in joint disease and intramedullary or extra-osseous extension in bone tumors.[27]

Another major advantage of MRI is that it is not associated with any radiation hazard. However, MRI has a longer acquisition time; hence it is more prone to motion artifacts. It is also more susceptible to artifacts from ferromagnetic substances, which can result in image distortion. The anatomical detail of the cortical bone is limited in MRI due to poor signals from the cortical bones. Also, MRI is not available as ubiquitously as CT and it is also more expensive.

4.3.2.4 3D Surface Scanners

A 3D scanner is a device akin to a camera that employs a blue or LASER light to capture multiple images and combine them to form a digital 3D model. The resultant (CAD) file is compatible with various software and can be used for 3D printing. This technology is a prompt and economical alternative to other imaging modalities described above. It is also non-invasive and more accurate than CT scanning, particularly for skin or teeth surface scans.

Depending upon the purpose, various types of 3D scanners are available. Desktop or intraoral 3D scanners are commonly used for 3D scanning of teeth and dental arches, aiding in the fabrication of guides for dental implants, guided root canal treatment, and the design of crowns and bridges, as well as for orthognathic surgeries. Conversely, handheld 3D scanners are used for 3D scanning of defective body parts, and those files are utilized to fabricate customized prostheses.

4.3.3 IMAGE ACQUISITION AND ITS PROTOCOL FOR PSI

Volumetric imaging on modern state-of-the-art CT and MRI machines has enabled fast and high-resolution acquisition of images with post-processing capabilities. Modern scanners permit the reformation of thin-slice 2D axial images into sagittal and coronal reformats and the generation of 3D volume-rendered images that can be rotated in any plane. These improved visualization techniques enable surgeons to make detailed plans regarding the surgical procedure on a patient-specific basis. All the medical images are stored in digital imaging and communications in medicine (DICOM) format, a standard data format to exchange medical images. For the high definition and minimal loss of actual anatomy data, there are set protocols for acquiring CT or MRI for various body parts. The following are the basic principles to be followed during CT scanning.

Non-fixed metallic dentures or jewelry in and near the region of interest should be removed. The patient's body part should be stabilized to remove motion artifacts. The patient should be scanned without gantry tilt (0° angulation) because gantry tilt compromises the quality of 3D reconstruction. Table height should be set so that the area of interest lies in the center of the scan field. Slice thickness should be 1 mm or less. Only the axial image dataset is acquired, which can be reformatted in any plane. A sharp reconstruction algorithm (bone or high resolution) should be used. The images are usually reconstructed using a 512 × 512 matrix.

For MR imaging, it is essential to optimize the protocol on a case-by-case basis because various factors alone or in combination can cause image degradation. MRI is more sensitive to motion; hence stabilization of the body part to be imaged is paramount. Higher field strength (3T) provides a higher signal-to-noise ratio (SNR) but it is also more susceptible to motion and metallic artifacts than 1.5T. The current limit of an in vivo MR resolution is 1 mm, and improvement in the resolution comes at the cost of imaging time and signal-to-noise ratio. Many anatomic regions are constrained in terms of the duration over which they can remain still; therefore, having longer acquisition times can result in motion artifacts and image blurring. In such cases, motion-compensating sequences are used.[43] Metallic bodies and air–tissue interfaces distort the local magnetic field and result in image degradation. Specialized sequences can help circumvent metallic artifacts such as multi-acquisition variable-resonance image combination (MAVRIC) from GE and slice-encoding for metal artifact correction (SEMAC) from Siemens. Spin-echo sequences are less susceptible to metallic artifacts than gradient-echo sequences.

4.3.4 ROLE OF IMAGING IN THE INTRAOPERATIVE PHASE

The image-guided computer-assisted static or dynamic navigation system helps the surgeon to place the implant precisely and effectively. This also avoids causing damage to critical anatomical structures.[44] An intraoperative radiographic assessment of the implant position aids in the identification and correction of the implant position, thus optimizing implant placement and mitigating additional revision surgery and anesthesia.[45] This is achieved using an x-ray C-arm or O-arm scanner, which captures 2D or 3D images, respectively. Intraoperative imaging also negates the need for imaging in the postoperative phase. However, intraoperative imaging has the disadvantage of high initial setup costs.[46,47] Another concern is radiation exposure to patients undergoing intraoperative imaging.

4.3.5 ROLE OF POST-IMPLANT IMAGING

Postoperative radiography, CT, and MRI are commonly used to evaluate the implant position and placement after surgery. They can also detect complications such as implant loosening, migration, and infection. A major challenge in postoperative imaging is the presence of metal artifacts in cases of metallic implants. These can cause significant image degradation on CT and MRI and render the images non-diagnostic. Fortunately, modern CT scanners and MRI machines have various metal artifact reduction (MAR) algorithms and software. In CT, these include iterative MAR (iMAR) from Siemens, SmartMAR from GE, and MAR for orthopedic implants (O-MAR) from Philips. As previously mentioned, metal artifact reduction sequences used in MRI include MAVRIC and SEMAC.

4.4 BASIC WORKFLOW OF THE PSI FROM IMAGING TO FABRICATION

The trend is leaning toward personalized medicine as surgeons aim to incentivize quality delivery to patients. Technological revolutions such as CAD/CAM have facilitated the adoption of the growth strategy of PSIs or PSPs and customized instrumentation. PSIs are commonly employed in the medical area, including orthopedic, craniomaxillofacial, dentistry, orthotics, and prosthetics.[38,48–50] With advancement comes the need for proficiency and familiarity by the treating surgeon rather than relying on technology to accomplish the task. In addition, to best design PSIs, collaborative meetings between the biomedical engineer and operating surgeon are essential. Various steps are involved in designing a PSI, and each step must be executed with caution to avoid the accumulation of errors. The following section outlines the technical requirements to manufacture PSIs. Subsequently, a broad overview of the steps implemented in the design, fabrication methods, scope, and challenges regarding PSIs is described. Figure 4.2 shows the main steps in the PSI fabrication workflow.

4.4.1 REQUIREMENTS

1. Raw data
   DICOM files—CT/MRI data
   3D image—.stl/.obj file exported from imaging software/scanner
2. Computer system
   Image processing and advanced CAD technology demand a high graphic resolution and storage space for intricate designing that has unleashed human ingenuity. Workstations are specially configured computers that can meet the technical computing requirements for image processing and CAD software. A computer system with 8GB RAM and a 2GB graphic card is the minimal requirement.
3. Software
   Different software is deployed during the workflow to combat complex functions and codes. The various steps, including image processing, modeling or designing, simulation, optimization, and personalization, each necessitates different software. Table 4.1 shows a list of commonly used imaging, CAD, and optimization software.
4. Hardware
   Haptic devices are force feedback tools that enable the designer to directly interact with the designed model and make changes in three-dimensional

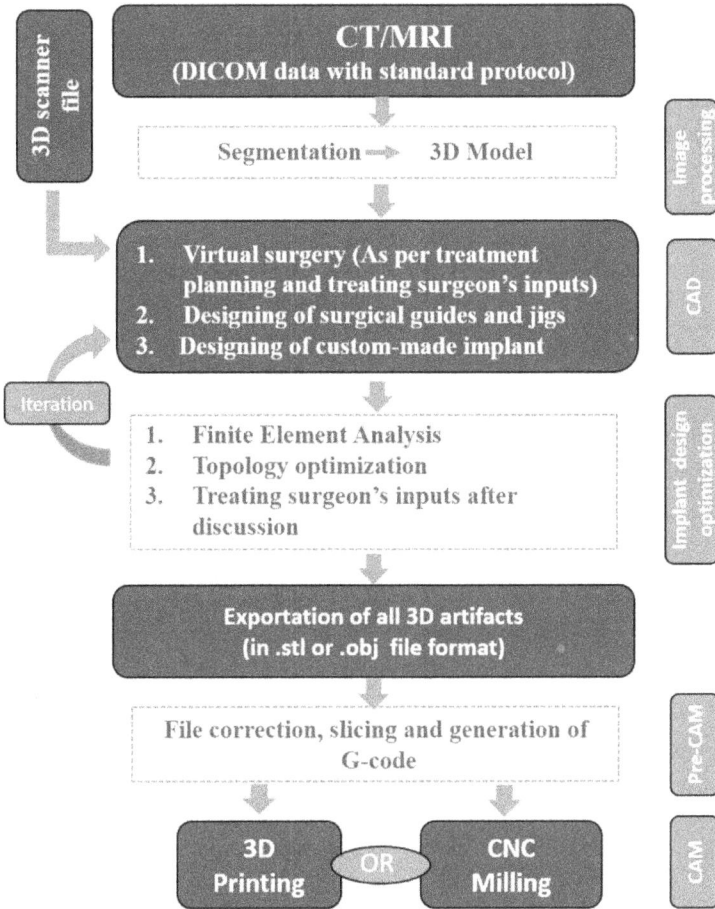

FIGURE 4.2 Schematic diagram showing a PSI fabrication workflow.

environments, thus enabling a critical interface between various CAD applications.[51] The surgeon can experience and verify the objects in a computer-generated environment.[52]

4.4.2 IMAGE PROCESSING

The design of PSIs begins with the acquisition of datasets. The selection of an appropriate imaging modality is vital to harness the advantages of the technological revolution. CT and MRI are the most commonly performed imaging technology when a PSI is planned. The choice of imaging depends on the advantages of spatial and contrast resolution. In simple terms, spatial resolution is the ability to differentiate between two bony landmarks and their relationship to surrounding structures. Contrast resolution is the ability to distinguish between regions of different densities

TABLE 4.1
Commonly Used Imaging, CAD, and Optimization Software

	Open	Commercial
Image processing	In Vesalius (CTI, Campinas, Brazil)	Materialise Mimics (Materialise NV, Leuven, Belgium)
	3D slicer (Surgical Planning Laboratory, Boston, MA)	3D-DOCTOR(Able Software Corp., Lexington, MA)
		Rhino 3D (Robert McNeel& Associates, Seattle, WA)
		Dolphin (Dolphin Imaging & Management Solutions, Chatsworth, CA)—orthognathic surgery
	ITK snap (Penn Image Computing and Science Laboratory, University of Pennsylvania, Philadelphia, PA)	IPS Gate (Implant Protesis Sistemas, Madrid, Spain)(KLS martin)
		Simpleware/ScanIP (Synopsys Inc., Mountain View, CA)
		Brainlab (Feldkirchen, Germany)
		D2P (3D Systems, Rock Hill, SC)
	OsiriX Lite (Pixmeo SARL, Geneva, Switzerland) (Mac users)	OsiriX (Pixmeo SARL, Geneva, Switzerland) Mac users
Designing	Meshmixer (Autodesk Inc., San Rafael, CA)	Proplan (Materialise, Leuven, Belgium)
	Blender (Blender Foundation, Amsterdam, the Netherlands)	3-matic (Materialise, Leuven, Belgium)
		Sculpt/Geomagic Freeform (3D Systems, Rock Hill, SC)
		Rhino 3D (Robert McNeel & Associates, Seattle, WA)
		SolidWorks (Dassault Systems, SolidWorks Corporation, Waltham, MA)
		Brainlab (Feldkirchen, Germany)
		CATIA V5 (Dassault Systems, Intrinsys)
FEA	—	ANSYS (Ansys Inc., Canonsburg, PA)
		Abaqus (Dassault Systèmes, Vélizy-Villacoublay, France)
		COMSOL Multiphysics (COMSOL Inc., Burlington, MA)
Topology optimization	—	Abaqus (Dassault Systèmes, Vélizy-Villacoublay, France)
		Geomagic (3D Systems, Rock Hill, SC)

or intensities. The raw data (CT, MRI) obtained is stored in the form of a DICOM format.

The DICOM format is the standard adopted for image data transfer. The images are stored in the form of slices. A 3D representation of a particular body part is obtained when the attributes embedded in each slice are convened along with pixel spacing.

Image processing involves converting these datasets into a virtual 3D image using the software. Factors such as the resolution of the original DICOM images, knowledge of the native anatomy, the expertise and techniques used for the segmentation of desired structures, and 3D reconstruction algorithms impact the final 3D model and sometimes contain artifacts. Specialized software such as Geomagic Wrap (Morrisville, NC) is utilized to smooth and filter the surface noise of 3D models. These files, which are stored in .stl or .obj format and are subsequently used for CAD of PSI. 3D images generated from the 3D scanner, can also be utilized directly in CAD of PSPs.

4.4.3 Computer-Aided Designing for PSI

After processing DICOM data into 3D images, the objects can be digitally manipulated by software. A virtual inspection of the 3D image provides details up to 0.1 mm precision. The virtual 3D image is manipulated, rendered, and analyzed after the inspection. Various software that serve the purpose are either open source or commercially available. The 3D models aid in identifying the anatomical landmarks of the target area and the design of the PSI. A virtual surgical procedure (VSP) includes digital cutting of bone and manipulation of segmented bony parts as per the treatment planning requirements of the individualized case. The location and orientation of the osteotomy plan are decided after consultation with the operating surgeon. Following the VSP based on the planned treatment strategy suggested by the treating surgeon, the virtual 3D model is optimal for designing PSIs and instrumentation. In oncology cases where the native anatomy is distorted, statistical shape modeling is introduced. Statistical shape modeling is a method that utilizes mathematical algorithms to effectively reconstruct the native anatomy, which enables an understanding of the geometric perturbations and biomechanics of the concerned area.[53] The CAD aspect involves designing a PSI and the auxiliary components required for the accurate transfer and execution of the plan. These include *models*, *instruments*, *guides*, *templates*, *jigs*, and *auxiliary tools.*

Tangible three-dimensional models of real anatomical sizes can be utilized as a presurgical tool for the preoperative execution of planned surgery and intraoperatively for guidance of the relevant anatomy. The 3D model fabricated after a VSP, referred to as the planned model, is intended for implant fit and the physical simulation of complex movements, and assesses the need for design optimization.[54] Patient-specific guides can serve various purposes such as bone cutting in the planned site, assisting in positioning drills, and the final position of the implant.[55] Custom templates can be designed to guide the rotational alignment of joints, such as in cases of total hip joint replacement. Patient-specific navigational templates and special jigs for virtual reality–assisted surgical simulation can be designed as per case requirements. Some special instruments such as personalized retractors or auxiliary tools specific to the patient for accessing complex regions can be designed. These can further help in holding the PSI in an accurate position and, in turn, reduce the overall surgical time and instrument load for sterilization.[56]

Molds are not an integral component of the PSI system. Instead, they are useful for the fabrication of a PSP for a specific need. The mold is used for the fabrication of silicone or poly methyl-methacrylate (PMMA)-based prostheses for the replacement of defects in various body parts, including the cranial, nasal, ear, and facial regions.[57,58] Following the design of a PSI and the auxiliary tools, all these files are exported in .stl or .obj format for FEA and topology optimization.

4.5 GENERAL DESIGNING STEPS AND PRINCIPLES

The general designing of a PSI begins with a discussion between the biomedical engineer and the surgeon during which the surgical steps and requirements are clarified. There is no specific set of rules for designing a PSI. Hence, the onus is on the surgeon to specify the anatomical location, dimensions, orientation, screw holes characteristics, and other entities such as guides and template requirements.

Many factors critically influence the design of a PSI, including the anatomical location, the involvement of a joint, the surgical approach, the magnitude of resection, advancement, setback, and angulation or alignment. Multiple iterations are required to design an optimal PSI. The following section describes the various principal considerations when designing a PSI for any region.

The design of components depends on the surgical access possible in the particular anatomical region. If, for example, the concerned site has limited exposure or is planned through a minimally invasive approach and the required dimension is larger than the achievable access to place the implant, the surgeon should consider the fabrication of multicomponents and assemble them later to achieve the desired outcome. The construct of a PSI is contingent upon its type, such as joint prostheses, which consist of multiple components, while the reconstructive type can comprise of one or two pieces depending on the complexity. In addition, the dimensions and extension also depend on the available bone stock that can support the implant placement and resist the arc of rotation and orientation to aid in a proper anatomical fit.

Design variables, such as the thickness of the implant, are dictated by their overlying soft tissue construct, the degree of augmentation required to achieve the desired contour in the concerned region, and also on potential fatigue stress.

Another important consideration during PSI design is the screw hole characteristics, e.g., number, location, diameter, and inter-hole distance. The number of screw holes depends on the extent of the implant and the minimum number required for stability. The factors determining the location include available bone buttress, proximity to vital structures, and stress fatigue. The diameter can be decided based on location and native bone characteristics. Inter-hole distance is determined by the strength of the remaining implant structures. The bone mapping process can determine the diameter and length of the screw for a particular hole. If the implant lies abutting vital structures, the screw hole is positioned to mitigate injury to vital structures. For example, the course of the inferior alveolar canal in the case of mandible reconstruction and the spinal cord in the case of spine reconstruction should be considered while designing.

The implant–bone contact (inner) surfaces should be rough enough to aid osseointegration. The exposed(outer) surface of the implant must be smooth to avoid stress accumulation and the formation of biofilms.[59,60] The space provision or crib features for placing bone grafts or flaps can be designed according to individual case requirements. Crib design aids in the percolation and extrusion of fluids in compartmental regions. Auxiliary design features such as chamfer edge, fillet, build volume, and support structures are done as fine-tuning and depend on the overall complexity of the PSI.

Once the design is finalized, visual superimposition and inspection are carried out in multiple 2D sections in the imaging software. Cross-fit is done on a 3D-printed model. The design of the cutting guide, templates, repositioning guide, or any specific tools is based on VSP and the final planned design of the PSI. The technological and educational differences between the surgeon and design engineer, each possessing special expertise and different considerations, can help in the optimal design of a PSI after multiple cyclic iterations. Further modification of the design of a PSI can be done on the basis of results obtained from FEA. The internal structure of a PSI can be enhanced by using a topology optimization process.

4.6   DESIGN OPTIMIZATION METHODS

4.6.1   Finite Element Modeling

Finite element modeling (FEM) is an advanced computer technology that facilitates the study of the biomechanical behavior of prosthetic components and various tissues. In simple terms, it is a software-based process that analyzes the reaction of the artifact to real-world forces, vibration, heat, fluid flow, or other physical effects. The process can determine whether a particular artifact will deform, fracture, deteriorate, or function in the way designed by assessing the stress distribution and strain throughout the components and host bone. In essence, FEA facilitates the design optimization of a PSI and the auxiliary fixtures.

To conduct an FEA, it is essential to have the mesh model of the patient native bone for PSI implantation and a mesh model of the PSI and auxiliary fixtures. Software such as Abaqus (Dassault Systèmes, France) is used to construct mesh models with hexahedral structures for implants and screws and tetrahedral elements for the bone. The element size is based on a mesh convergence study, which considers the complexity of the bone and implant structure.[61] Following mesh model creation, the properties of the bone and implant are established. The native bone is usually considered to be isotropic and linearly elastic for the magnitude of loads. The value of the implant properties differ depending on the material used to fabricate the PSI. Before computational simulation, the properties of the materials, boundary conditions, loading, and bone support must be properly defined. Boundary conditions define the bone position and fixing the bone end during muscle load, assigning the muscle load and its direction. After setting up the parameters, the FEA outcome can be expressed in terms of deformation (strain; in mu) and stress distribution around the screw or throughout the implant (von Mises; in MPa).[62] Additionally, micromovements can be depicted (in mm) during load. These results facilitate design optimization in areas of stress and strain.

4.6.2   Topology Optimization

After finalizing the PSI design, topology optimizationis crucial for further improvisation of the structural design. Topology optimization includes gyroid lattice structures to reduce the weight of mechanical structures and help in osseointegration and angiogenesis.[63,64] This will enhance the integration of the implant with the surrounding bone and tissues.[65,66] The robustness of a PSI following the incorporation of a lattice structure is checked before implantation. TO is not usually possible with CNC-made implants.

4.7 ROLE OF NEW TECHNOLOGY IN PSI DESIGNING

Technological advancement has aided in 3D visualization, preoperative planning, and the manufacture of PSIs. All these processes are advantageous to the surgeon but are labor intensive and time-consuming. The surgeon should be familiar with the steadfast revolution of technology. For instance, precise segmentation during image processing, VSP, intricate designing of implants, and optimization are time-consuming and arduous. Recently, an artificial intelligence (AI)-based system has played a pervasive role and enabled the execution of such complex tasks with a single click of a button. The employment of AI in the health sector could unfold human ingenuity, limit cognitive bias, and, in turn, reduce errors. In the current era of personalized medicine, the integration of AI in designing PSIs is not unexplored.

The technological evolution in medical datasets, the sophistication of neural networks, and computing powers allow for the rapid development of artificial intelligence in the medical field. AI-integrated CAD, called model-based reasoning, helps span the technology across different levels, enabling multifactor analysis of a single model as dictated by the automated AI algorithm. Essentially, the incorporation of AI in CAD streamlines designing, archives knowledge, facilitates change without human intervention, and reduces overall time. Machine learning (ML),a subset of AI algorithms, can analyze the data in DICOM format, auto-segment, and create a three-dimensional model of the interested region. These 3D models are the fundamental basis for the design of custom implants tailored to the unique needs of the patient. For example, a convolutional neural network (CNN) was utilized for image processing in total hip arthroplasty and found better outcomes with a PSI. The drawback of this method is the need for numerous datasets to train deeplearning neural networks. However, with an increase in datasets, accuracy can be improved.

The assumptions that human body parts are almost symmetrical around the midsagittal plane are the main backbone for the design of implants based on Boolean operations on pristine body parts. An image processing algorithm helps to identify anatomical landmarks such as the midline sagittal plane. Proposed PSI fabrication techniques involve a combination of image processing algorithms, computational geometry, interpolation methods, and optimization. Some utilize deep learning networks based on volumetric completion for cranioplasty designing. Literature exists regarding the image ray approach used to identify the inner and outer table of the skull, subtraction methods to design the cranioplast, and pixel encoding to precisely align the implant, which is vital for generating an accurate PSI.[67] In the future, AI will design implants using point-and-click operations. The design of patient-specific spinal screws has been attempted based on artificial neural networks from the data generated by FEA.[68] The design optimization of femur implants can be carried out with ANN surrogate models.[69] On the contrary, the complexity involved in FEA hampers its automation. However, all these bottleneck complexities will be addressed by evolving technologies.

Along with AI, augmented reality (AR) and virtual reality also have a role in CAD design. Virtual reality can aid in a comprehensive understanding of the design through immersive simulation at spatially designed locations, while augmented reality allows the merging of real-world and virtual data.[70] The incorporation of these advanced technologies will greatly reduce the lead time. However, they need rigorous validation before adoption. The newer technologies are attractive and constantly evolving, but have certain limitations.[70]

4.8 PSI FABRICATION PROCESS

CNC milling is considered an artisanal method for fabricating CMIs. However, this method has many limitations and drawbacks, such as difficulty in manufacturing complex, anatomically precise implants, lengthy turnaround times, increased cost, and the requirement of skilled labor. Additionally, wastage of material is inevitable. Consequently, CNC is suitable for a small and more straightforward implant fabrication. With the advent of novel software for design optimization, improved biomaterials, and 3D printing technology, the whole CMI process has transformed. Even intricate and large implants can be fabricated with minimal material waste using 3D printing technology such as selective laser melting (SLM) or electron beam melting (EBM). Nevertheless, 3D printing has its drawbacks, including the need for extensive post-processing and questionable strength. Both technologies have potential benefits and limits; hence, the choice depends on the individual case and accessibility.

4.8.1 PREMANUFACTURING REQUIREMENTS

The CAD files of a PSI and other auxiliary entities such as 3D models (anatomy and planned), surgical cutting, positioning guides or templates, molds, and instrumentation are prime requirements. The technical needs include the software and biomaterial for PSI fabrication. The manufacturing material of choice for PSI and auxiliary components is enumerated in Table 4.2. Specialized software is necessary to examine the .stl file for any errors before proceeding with manufacturing. 3D printers and CNC machines cannot read .stl files; therefore, they need a special slicing code called G-code for final product fabrication. This is accomplished with the use of CAM software.

Various biomaterials, such as titanium and its alloy,[71] stainless steel, chromium-nickel-cobalt alloy, zirconium, and polymers[72] including silicon (polymerized dimethylsiloxane), poly-ether-ether-ketone (PEEK),[73] polyether-ketone-ketone (PEKK),[74] ultra-high molecular weight polyethylene (UHMWPE), and poly methylmethacrylate, can be utilized to manufacture different components of the PSI system. Powder-form materials are utilized for 3D printing by selective laser sintering (SLS), multi-jet fusion (MJF), SLM, EBM, plastic loops for fused deposition modeling

TABLE 4.2

Various Types of Manufacturing Technologies and Biomaterials for PSI and Auxiliary Components

Artifacts	Type of manufacturing		Biomaterial
Anatomy model, planned model, and interim PSI	Additive (3D printing)	FDM SLA SLS	ABS/PLA Resin Nylon
Surgical guide, positioning guide, and custom tool/instrument	Additive (3D printing)	SLS SLM/EBM	Nylon, titanium and its alloy, stainless steel, chromium-nickel-cobalt alloy, etc.
	Subtractive (CNC)	5 or greater than axis CNC	Stainless steel, chromium-nickel-cobalt alloy, etc.
Direct PSI (cranioplast, other bone reconstructive part including joint component)	Additive (3D printing)	SLM/EBM	Titanium and its alloy, stainless steel, etc.
	Subtractive (CNC)	5 or greater than axis CNC	PEEK, UHMWPE, zirconium, titanium and its alloy, stainless steel, etc.
Mold	Additive (3D printing)	SLS, FDM	Nylon (polyamide) and acrylonitrile butadiene styrene (ABS)
Mold-based PSP (craniofacial and other soft tissue prostheses)	Using a 3D-printed mold	Manual procedure	PMMA and silicon

(FDM), liquid resin for stereolithography (SLA) and digital light processing (DLP), and metal or plastic blocks for CNC machining. For instance, guides or templates can be printed using either metal (SLM/EBM), plastic (SLS), resin (SLA), or CNC milling. Metal-based guides are best in terms of compact design, better fit, and low risk of grinding during use. A 3D-printed PSI might occasionally require add-on CNC milling for final finish and surface characterization. Nevertheless, the choice of technology, whether additive or subtractive, depends on availability, feasibility, type of prosthesis, and affordability.

4.8.2 CAM FOR PSI

In the early days, the fabrication of a PSI was cumbersome and almost impossible in a complex situation. The advent of multi-axis CNC and 3D printing, along with advanced software, for optimized designing, propelled the manufacture of PSIs to new heights. Nowadays, a PSI has become an essential part of reconstructive surgery as a component of personalized medicine. Initially, CNC was the only solution inlimited cases, but the subsequent introduction of improved 3D printing technology changed the entire gamut of PSI fabrication.

4.8.2.1 CNC Milling

CNC machining is a subtractive manufacturing process that involves the removal of material from a solid block to fabricate physical artifacts. The intricacy of the work done by CNC machines depends on the number of axes. More than three axes are required to manufacture a medical implant. CNC machines with five axes or more are preferred for plastic and metal milling. Multi-axes CNC machines greatly reduce the time needed for manufacturing and misalignment errors and are beneficial in fabricating complex components.

Although CNC machines offer many benefits over 3D printing, there are certain limitations when it comes to the fabrication of complex structures, such as generating thin walls and undercuts and the inability to create hollow or lattice structures. In addition, CNC is time-consuming, costly, and results in the wastage of material. Despite this, whenever feasible, such as a minimally complex artifact, no cost considerations, and easy access to a facility, a CNC-made PSI should be considered. These CNC-made PSIs exhibit relatively better material properties and require minimal post-processing. Currently, blocks of stainless steel, titanium, zirconium, Cr-Co-Ni alloy, and PEEK are used for CNC milling.

4.8.2.2 3D Printing

3D printing is an example of additive manufacturing. It involves the layer-by-layer addition of material to manufacture a physical artifact. The type of 3D printing technology employed depends on the component of the PSI. For anatomical 3D models with pathology or virtually planned models, FDM, SLA, and SLS techniques may be used depending on the availability of services and the precision requirements for individual case scenarios. SLA- and SLS-printed models are considered more accurate than FDM technology.

Commonly used 3D printing technology for plastic artifacts includes fused deposition modeling, SLS, and MJF. Metal artifacts are printed using SLM and EBM, while resin artifacts are printed using SLA and DLP technology. The surgical guide, template, and customized instrument can be printed using plastic or resin materials, while PSIs and instrumentation can be manufactured using metal. A detailed description of individualized 3D printing technology is beyond the scope of this chapter.

3D printing offers several benefits compared to CNC, including faster production times, lower costs, and the ability to print hollow artifacts. Tremendous improvements in technology, materials, and research are leading to growing popularity in the medical field.

4.8.3 Post-CAM Processing

Post-CAM processing includes a series of steps before the final delivery of the artifact and implantation. This process encompasses procedures such as post-curing, post-heat treatment, sandblasting, refining screw holes, and final cleaning and finishing, achieving a polished appearance.

Biomechanical testing includes multiple bench tests to check the overall quality and strength of the implant. For example, fatigue testing is based on the simulation of actual biomechanical loads, e.g., 250,000 times for a mandible PSI, which is equivalent to the load sustained after six months following surgery.[75]

The sterilization method is crucial for all components of a PSI before implantation in the human body. The type of sterilization method depends on the material used for a particular component and available amenities. Low-temperature sterilization methods such as gas plasma or ethylene oxide (EtO) can be used for plastic-made (acrylonitrile butadiene styrene [ABS] or polylactic acid [PLA] or UHMWPE or PEEK, etc.) artifacts such as models, surgical guides, and implants. Usually, metal parts are sterilized using the Gamma radiation method.

4.9   PSI-RELATED CHALLENGES AND LIMITATIONS

With technological advancements come limitations and challenges. The design process is time-consuming as the planning is didactic and requires multiple steps. It has a steep learning curve, and software limitations exacerbate the difficulties. Manufacturing can be costly with limited services, and legal compliance can further increase the turnaround time and hinder the routine use of a PSI. Although emerging technologies such as navigation, robotics, and mixed reality have aided in the accurate implantation of a PSI, some persistent micro displacements may lead to metallic wear.

4.10   SUMMARY

The advent of CAD/CAM technology has allowed more personalized delivery of medical healthcare. PSIs have become a biomechanical alternative in medical specialties such as maxillofacial surgery, orthopedic surgery, neurosurgery, dentistry, orthotics, and prosthetics. The selection of an appropriate imaging modality depends on the defect or region requiring restoration or rehabilitation. CT is the preferred imaging modality for implants replacing hard tissue defects. MRI is considered for soft tissue defects or as an additional detail besides CT. 3D scanners are used to scan the surface of teeth and for soft tissue augmentation implants. Advanced imaging modalities, along with CAD/CAM, facilitate the design of complex implants that can restore functional and aesthetic limitations. The design process is didactic, and multiple cyclic iterations are essential after discussions between the surgeon and engineer. On completion of the design, manufacturing can be accomplished via CNC machining, 3D printing, or hybrid methods, depending on the individual requirement. Further advancements in software and the incorporation of artificial intelligence networks have the potential to streamline the process, reducing cost and overall turnover time.

REFERENCES

1. Maniar RN, Singhi T. Patient-specific implants: Scope for the future. *Curr Rev Musculoskelet Med* 2014 Jun; 7(2): 125–30.

2. Owusu JA, Boahene K. Update of patient-specific maxillofacial implant. *Curr Opin Otolaryngol Head Neck Surg* 2015 Aug; 23(4): 261–4.
3. Binder WJ. Custom-designed facial implants. *Facial Plast Surg Clin North Am* 2008 Feb; 16(1): 133–46, vii.
4. Goldsmith D, Horowitz A, Orentlicher G. Facial skeletal augmentation using custom facial implants. *Atlas Oral Maxillofac Surg Clin North Am* 2012 Mar; 20(1): 119–34.
5. Binder WJ, Bloom DC. The use of custom-designed midfacial and submalar implants in the treatment of facial wasting syndrome. *Arch Facial Plast Surg* 2004; 6(6): 394–7.
6. Scolozzi P, Martinez A, Jaques B. Complex orbito-fronto-temporal reconstruction using computer-designed PEEK implant. *J Craniofac Surg* 2007 Jan; 18(1): 224–8.
7. Alasseri N, Alasraj A. Patient-specific implants for maxillofacial defects: Challenges and solutions. *Maxillofac Plast Reconstr Surg* 2020; 42(1): 15.
8. Sbordone C, Toti P, Guidetti F, Califano L, Pannone G, Sbordone L. Volumetric changes after sinus augmentation using blocks of autogenous iliac bone or freeze-dried allogeneic bone. A non-randomized study. *J Craniomaxillofac Surg* 2014 Mar; 42(2): 113–8.
9. Kim MM, Boahene KDO, Byrne PJ. Use of customized polyetheretherketone (PEEK) implants in the reconstruction of complex maxillofacial defects. *Arch Facial Plast Surg* 2009; 11(1): 53–7.
10. Du R, Su Y-X, Yan Y, Choi W, Yang W, Zhang C, et al. A systematic approach for making 3D-printed patient-specific implants for craniomaxillofacial reconstruction. *Engineering* 2020 Nov 1; 6(11): 1291–301.
11. Binder WJ, Kaye A. Reconstruction of post-traumatic and congenital facial deformities with three-dimensional computer-assisted custom-designed implants. *Plast Reconstr Surg* 1994 Nov; 94(6): 775–7.
12. Kozakiewicz M, Elgalal M, Walkowiak B, Stefanczyk L. Technical concept of patient-specific, ultrahigh molecular weight polyethylene orbital wall implant. *J Craniomaxillofac Surg* 2013 Jun; 41(4): 282–90.
13. Gander T, Essig H, Metzler P, Lindhorst D, Dubois L, Rücker M, et al. Patient specific implants (PSI) in reconstruction of orbital floor and wall fractures. *J Craniomaxillofac Surg* 2015 Jan; 43(1): 126–30.
14. Mertens C, Löwenheim H, Hoffmann J. Image Data Based Reconstruction of the Midface. using a patient-specific implant in combination with a vascularized osteomyocutaneous scapular flap. *J Cranio-maxillo-facial Surg.* 2013 Apr; 41(3): 219–25.
15. Patel N, Mel A, Patel P, Fakkhruddin A, Gupta S. A Novel Method to Rehabilitate Post-mucormycosis maxillectomy Defect by Using Patient-Specific zygoma Implant. *J Maxillofac Oral Surg* 2023 Feb; 22(Suppl 1): 1–6.
16. Mommaerts MY. Evolutionary steps in the design and biofunctionalization of the additively manufactured sub-periosteal jaw implant "AMSJI" for the maxilla. *Int J Oral Maxillofac Surg* 2019 Jan; 48(1): 108–14.
17. Vosselman N, Merema BJ, Schepman KP, Raghoebar GM. Patient-specific sub-periosteal zygoma implant for prosthetic rehabilitation of large maxillary defects after oncological resection. *Int J Oral Maxillofac Surg* 2019 Jan; 48(1): 115–7.
18. Rotaru H, Schumacher R, Kim S-G, Dinu C. Selective laser melted titanium implants: A new technique for the reconstruction of extensive zygomatic complex defects. *Maxillofac Plast Reconstr Surg* 2015; 37(1): 1.
19. Essig H, Lindhorst D, Gander T, Schumann P, Könü D, Altermatt S, et al. Patient-specific biodegradable implant in pediatric craniofacial surgery. *J Craniomaxillofac Surg* 2017 Feb; 45(2): 216–22.

20. Li B, Wei H, Jiang T, Qian Y, Zhang T, Yu H, et al. Randomized clinical trial of the accuracy of patient-specific implants versus CAD/CAM splints in orthognathic surgery. *Plast Reconstr Surg* 2021 Nov; 148(5): 1101–10.

21. Zhou L, He L, Shang H, Liu G, Zhao J, Liu Y. Correction of hemifacial microsomia with the help of mirror imaging and a rapid prototyping technique: Case report. *Br J Oral Maxillofac Surg* 2009 Sep; 47(6): 486–8.

22. Kärkkäinen M, Wilkman T, Mesimäki K, Snäll J. Primary reconstruction of orbital fractures using patient-specific titanium milled implants: The Helsinki protocol. *Br J Oral Maxillofac Surg* 2018 Nov; 56(9): 791–6.

23. Raisian S, Fallahi HR, Khiabani KS, Heidarizadeh M, Azdoo S. Customized Titanium Mesh Based on the 3D Printed Model vs. Manual Intraoperative Bending of Titanium Mesh for Reconstructing of Orbital Bone Fracture: A Randomized Clinical Trial. *Rev Recent Clin Trials* 2017; 12(3): 154–8.

24. Burkhard JPM, Koba S, Schlittler F, Iizuka T, Schaller B. Clinical results of two different three-dimensional titanium plates in the treatment of condylar neck and base fractures: A retrospective study. *J Craniomaxillofac Surg* 2020; 48(8): 756–64.

25. Mustafa SF, Evans PL, Bocca A, Patton DW, Sugar AW, Baxter PW. Customized titanium reconstruction of post-traumatic orbital wall defects: A review of 22 cases. *Int J Oral Maxillofac Surg* 2011 Dec; 40(12): 1357–62.

26. Wong KC. 3D-printed patient-specific applications in orthopedics. *Orthop Res Rev* 2016; 8: 57–66.

27. Kilian D, Sembdner P, Bretschneider H, Ahlfeld T, Mika L, Lützner J, et al. 3D printing of patient-specific implants for osteochondral defects: Workflow for an MRI-guided zonal design. *Bio-Des Manuf* 2021; 4(4): 818–32.

28. Senkoylu A, Daldal I, Cetinkaya M. 3D printing and spine surgery. *J Orthop Surg (Hong Kong)* 2020; 28(2): 2309499020927081.

29. Kurtz SM, Devine JN. PEEK biomaterials in trauma, orthopedic, and spinal implants. *Biomaterials* 2007 Nov; 28(32): 4845–69.

30. Yang X, Wan W, Gong H, Xiao J. Application of individualized 3D-printed artificial vertebral body for cervicothoracic reconstruction in a Six-Level recurrent chordoma. *Turk Neurosurg* 2020; 30(1): 149–55.

31. Fan H, Fu J, Li X, Pei Y, Li X, Pei G, et al. Implantation of customized 3-D printed titanium prosthesis in limb salvage surgery: A case series and review of the literature. *World J Surg Oncol* 2015 Nov; 13: 308.

32. Wong KC, Kumta SM, Geel NV, Demol J. One-step reconstruction with a 3D-printed, biomechanically evaluated custom implant after complex pelvic tumor resection. *Comput Aid Surg* 2015; 20(1): 14–23.

33. Chatain GP, Finn M. Compassionate use of a custom 3D-printed sacral implant for revision of failing sacrectomy: Case report. *J Neurosurg Spine* 2020; 33: 513–18.

34. Lador R, Regev G, Salame K, Khashan M, Lidar Z. Use of 3-dimensional printing technology in complex spine surgeries. *World Neurosurg* 2020 Jan; 133: e327–41.

35. Kondaka S, Singh VD, Vadlamudi C, Bathala LR. Prosthetic rehabilitation of untailored defects using patient-specific implants. *Dental Res J.* 2022; 19: 83.

36. Barrios-Muriel J, Romero-Sánchez F, Alonso-Sánchez FJ, Rodríguez Salgado D. Advances in orthotic and prosthetic manufacturing: A technology review. *Mater (Basel, Switzerland)* 2020 Jan; 13(2): 295.

37. Faustini MC, Neptune RR, Crawford RH, Stanhope SJ. Manufacture of Passive Dynamic ankle-foot orthoses using selective laser sintering. *IEEE Trans Bio Med Eng* 2008 Feb; 55(2 Pt 1): 784–90.

38. Mavroidis C, Ranky RG, Sivak ML, Patritti BL, DiPisa J, Caddle A, et al. Patient specific ankle-foot orthoses using rapid prototyping. *J Neuroeng Rehabil* 2011; 8(1): 1.

39. Gupta S, Patil N, Solanki J, Singh R, Laller S. Oral implant imaging: A review. *Malays J Med Sci* 2015; 22(3): 7–17.
40. Huppertz A, Radmer S, Wagner M, Roessler T, Hamm B, Sparmann M. Computed tomography for pre-operative planning in total hip arthroplasty: What radiologists need to know. *Skelet Radiol* 2014 Aug; 43(8): 1041–51.
41. Lee T-Y, Chhem RK. Impact of new technologies on dose reduction in CT. *Eur J Radiol* 2010 Oct; 76(1): 28–35.
42. Kumar M, Shanavas M, Sidappa A, Kiran M. Cone beam computed tomography - Know its secrets. *J Int Oral Heal JIOH* 2015 Feb; 7(2): 64–8.
43. Ripley B, Levin D, Kelil T, Hermsen JL, Kim S, Maki JH, et al. 3D printing from MRI Data: Harnessing strengths and minimizing weaknesses. *J Magn Reson Imaging* 2017 Mar; 45(3): 635–45.
44. Kalaivani G, Balaji VR, Manikandan D, Rohini G. Expectation and reality of guided implant surgery protocol using computer-assisted static and dynamic navigation system at present scenario: Evidence-based literature review. *J Indian Soc Periodontol* 2020; 24(5): 398–408.
45. Nguyen E, Lockyer J, Erasmus J, Lim C. Improved outcomes of orbital reconstruction with intraoperative imaging and rapid prototyping. *J Oral Maxillofac Surg* 2019 Jun; 77(6): 1211–7.
46. Graham DO, Lim CGT, Coghlan P, Erasmus J. A literature review of rapid prototyping and patient specific implants for the treatment of orbital fractures. *Craniomaxillofac Trauma Reconstr* 2022 Mar; 15(1): 83–9.
47. Shaye DA, Tollefson TT, Strong EB. Use of intraoperative computed tomography for maxillofacial reconstructive surgery. *JAMA Facial Plast Surg* 2015; 17(2): 113–9.
48. Memon AR, Wang E, Hu J, Egger J, Chen X. A review on computer-aided design and manufacturing of patient-specific maxillofacial implants. *Expert Rev Med Devices* 2020 Apr; 17(4): 345–56.
49. Haglin JM, Eltorai AEM, Gil JA, Marcaccio SE, Botero-Hincapie J, Daniels AH. Patient-specific orthopaedic implants. *Orthop Surg* 2016 Nov; 8(4): 417–24.
50. Huang MF, Alfi D, Alfi J, Huang AT. The use of patient-specific implants in oral and maxillofacial surgery. *Oral Maxillofac Surg Clin North Am* 2019 Nov; 31(4): 593–600.
51. Liu X, Dodds G, McCartney J, Hinds BK. Virtual DesignWorks—Designing 3D CAD models via haptic interaction. *Comput Des* 2004; 36(12): 1129–40.
52. Escobar-Castillejos D, Noguez J, Neri L, Magana A, Benes B. A review of simulators with haptic devices for medical training. *J Med Syst* 2016; 40(4): 104.
53. Ambellan F, Lamecker H, von Tycowicz C, Zachow S. Statistical shape models: Understanding and mastering variation in anatomy. *Adv Exp Med Biol.* 2019; 1156: 67–84.
54. Parthasarathy J. 3D modeling, custom implants and its future perspectives in craniofacial surgery. *Ann Maxillofac Surg* 2014 Jan; 4(1): 9–18.
55. Meng M, Wang J, Sun T, Zhang W, Zhang J, Shu L, et al. Clinical applications and prospects of 3D printing guide templates in orthopaedics. *J Orthop Translat* 2022; 34: 22–41.
56. Hafez MA, Moholkar K. Patient-specific instruments: Advantages and pitfalls. *SICOT J* 2017; 3: 66.
57. Neto R, Costa-Ferreira A, Leal N, Machado M, Reis A. An engineering-based approach for design and fabrication of a customized nasal prosthesis. *Prosthet Orthot Int* 2015; 39(5): 422–8.
58. Lal B, Ghosh M, Agarwal B, Gupta D, Roychoudhury A. A novel economically viable solution for 3D printing-assisted cranioplast fabrication. *Br J Neurosurg* 2020 Jun; 34(3): 280–3.

59. Kligman S, Ren Z, Chung C-H, Perillo MA, Chang Y-C, Koo H, et al. The impact of dental implant surface modifications on osseointegration and biofilm formation. *J Clin Med* 2021; 10(8): 1641.

60. Bandyopadhyay A, Mitra I, Goodman SB, Kumar M, Bose S. Improving biocompatibility for next generation of metallic implants. *Prog Mater Sci* 2023; 133: 101053.

61. Rodrigues YL, Mathew MT, Mercuri LG, da Silva JSP, Henriques B, Souza JCM. Biomechanical simulation of temporomandibular joint replacement (TMJR) devices: A scoping review of the finite element method. *Int J Oral Maxillofac Surg* 2018; 47(8): 1032–42.

62. Lee C-H, Mukundan A, Chang S-C, Wang Y-L, Lu S-H, Huang Y-C, et al. Comparative analysis of stress and deformation between one-fenced and three-fenced dental implants using finite element analysis. *J Clin Med* 2021 Sep; 10(17): 3986.

63. Mahmoud D, Elbestawi MA. Lattice structures and functionally graded materials applications in additive manufacturing of orthopedic implants: A review. *J. Manuf. Mater.* 2017; 1(2): 13.

64. Lin C-L, Wang Y-T, Chang C-M, Wu C-H, Tsai W-H. Design criteria for patient-specific mandibular continuity defect reconstructed implant with lightweight structure using weighted topology optimization and validated with biomechanical fatigue testing. *Int J Bioprinting* 2022; 8(1): 437.

65. Bobbert FSL, Zadpoor AA. Effects of bone substitute architecture and surface properties on cell response, angiogenesis, and structure of new bone. *J Mater Chem B* 2017 Aug; 5(31): 6175–92.

66. Wang Z, Wang C, Li C, Qin Y, Zhong L, Chen B, et al. Review. *J Alloys Compd.* 2017; 717(C): 271–85.

67. Venugopal V, Ghalsasi O, McConaha M, Xu A, Forbes J, Anand S. Image processing-based method for automatic design of patient-specific cranial implant for additive manufacturing. *Procedia Manuf* 2021; 53: 375–86.

68. Biswas JK, Dey S, Karmakar SK, Roychowdhury A, Datta S. Design of patient specific spinal implant (pedicle screw fixation) using FE analysis and soft computing techniques. *Curr Imaging* 2020; 16(4): 371–82.

69. Chatterjee S, Dey S, Majumder S, RoyChowdhury A, Datta S. Computational intelligence based design of implant for varying bone conditions. *Int J Numer Method Biomed Eng* 2019; 35(6): e3191.

70. Ghaednia H, Fourman MS, Lans A, Detels K, Dijkstra H, Lloyd S, et al. Augmented and virtual reality in spine surgery, current applications and future potentials. *Spine J* 2021 Oct; 21(10): 1617–25.

71. Yang J, Liu C, Sun H, Liu Y, Liu Z, Zhang D, et al. The progress in titanium alloys used as biomedical implants: From the view of reactive oxygen species. *Front Bioeng Biotechnol* 2022; 10: 1092916.

72. Abdo Filho RCC, Oliveira TM, Lourenço Neto N, Gurgel C, Abdo RCC. Reconstruction of bony facial contour deficiencies with polymethylmethacrylate implants: Case report. *J Appl Oral Sci.* 2011 Aug; 19(4): 426–30.

73. Nieminen T, Kallela I, Wuolijoki E, Kainulainen H, Hiidenheimo I, Rantala I. Amorphous and crystalline polyetheretherketone: Mechanical properties and tissue reactions during a 3-year follow-up. *J Biomed Mater Res A* 2008 Feb; 84(2): 377–83.

74. Alqurashi H, Khurshid Z, Syed AUY, Rashid Habib S, Rokaya D, Zafar MS. Polyetherketoneketone (PEKK): An emerging biomaterial for oral implants and dental prostheses. *J Adv Res* 2021; 28: 87–95.

75. Schupp W, Arzdorf M, Linke B, Gutwald R. Biomechanical testing of different osteosynthesis systems for segmental resection of the mandible. *J Oral Maxillofac Surg* 2007 May; 65(5): 924–30. doi: 10.1016/j.joms.2006.06.306.

5 Finite Element Analysis of Medical Implants

Ayush Srivastav, Ravi Kumar Dwivedi,
John Ashutosh Santoshi, and Prateek Behera

5.1 INTRODUCTION

Medical implants are synthetically created devices that are inserted into the human body for medical purposes. They are used to support organs and tissues, replace body parts such as hip and knee joints, distribute medication for pain treatment, and monitor or control bodily processes such as the heart rate. Some medical implants are inert and are only meant to offer support, e.g., surgical stents and screws, whereas some implants are active, i.e., they communicate with the body, e.g., pacemakers which deliver shocks of electricity in response to variations in the heartbeat.

The focus of this chapter is on inert medical implants and their design, especially from the orthopedic and dentistry usage point of view. Surgery for the reduction and retention (fixation) of fractures needs specifically designed implants.

Finite element analysis (FEA) is used for the computational analysis and validation of these implant designs before any type of human trials. Feedback from FEA results about implant deformation or the movement of bone fragments helps in redesigning the implants. This ensures minimum failure cases in real life by redesigning the implants for survival in the maximum number of cases before proceeding to human trials.

In order to understand how FEA is useful, a scaphoid waist fracture and the implant used for its fixation were selected for detailed analysis in this chapter.

5.1.1 SCAPHOID BONE

The scaphoid bone is one of the eight carpal bones (scaphoid, lunate, capitate, hamate triquetrum, pisiform, trapezium, trapezoid) in the human wrist. It is situated distal to the radius on the radial (thumb) side of the wrist, as shown in Figure 5.1. It is approximately the size and shape of a medium cashew nut, as shown in Figure 5.2.

The scaphoid bone is prone to fractures if a person falls onto their outstretched hand and is infamously difficult to self-heal due to the peculiar pattern of the blood supply in the anatomical snuffbox region where the vessels supplying the proximal part course through the body of the scaphoid after entering the bone from the distal

DOI: 10.1201/9781003375098-5

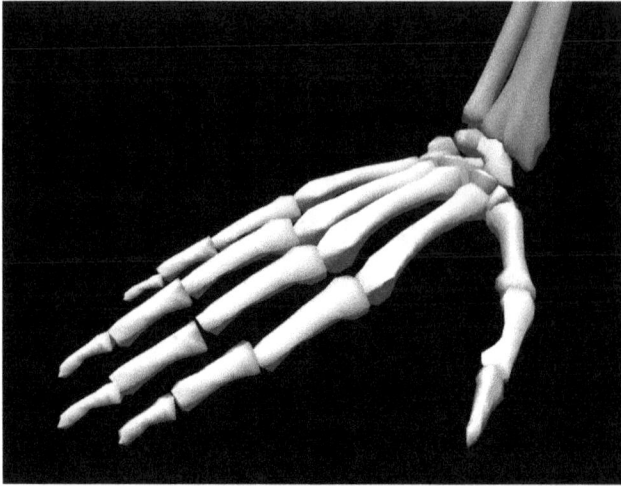

FIGURE 5.1   Hand skeleton [1].

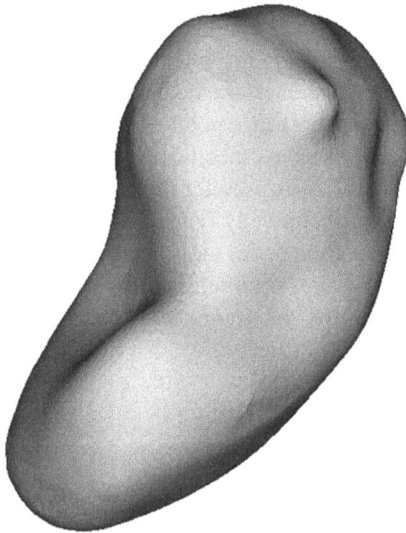

FIGURE 5.2   Scaphoid bone.

end. Implants such as Kirschner wire (K-wire) and headless screws are used in the fixation of the scaphoid.

5.1.2   FRACTURES

Scaphoid fractures [2] are generally classified as either displaced or non-displaced.

A non-displaced scaphoid fracture means that the bone fragments (pieces) have not shifted from their normal position, and at times the fracture may not even be visible on an x-ray image. These types of fractures are usually treated using hand casts, which immobilize the hand for a few months to allow bone recuperation.

On the contrary, the fragments of a displaced scaphoid fracture shift out of their normal position. Furthermore, scaphoid fractures are also classified according to the severity of the displacement of the bone fragments. The Mayo classification of scaphoid fractures divides scaphoid fractures into three types according to the anatomic location of the fracture line: middle or waist (70%), distal (20%), and proximal (10%) [3] (Figures 5.3–5.5). The shape of the scaphoid bone and its mechanism of fracture [4] are believed to be the cause of 70% of scaphoid fractures occurring at the middle (waist).

5.2 MEDICAL IMPLANTS

5.2.1 K-WIRE

K-wire [5] is a thin metallic wire made of stainless steel that is used to stabilize fractured bone fragments. This wire can be drilled through the bone to hold the fragments in place; it can be placed percutaneously or buried beneath the skin. K-wires are usually used for temporary fixation during operations and are removed after definitive fixation. However, if the fracture fragments are small, the K-wire can be used for definitive fixation (e.g., wrist fractures).

A rendered image of a K-wire is shown in Figure 5.6. The K-wire features two extremely sharp edges and is available in different lengths and diameters. K-wires

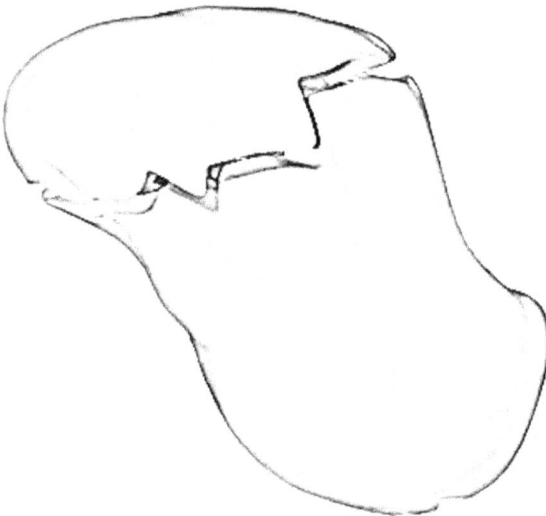

FIGURE 5.3 Distal fracture in a scaphoid bone.

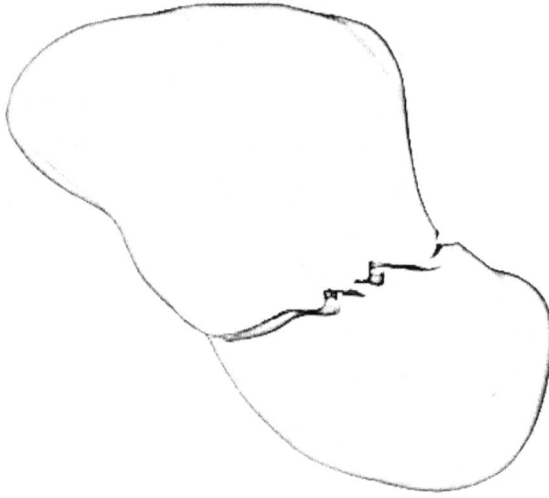

FIGURE 5.4   Waist (middle) fracture in a scaphoid bone.

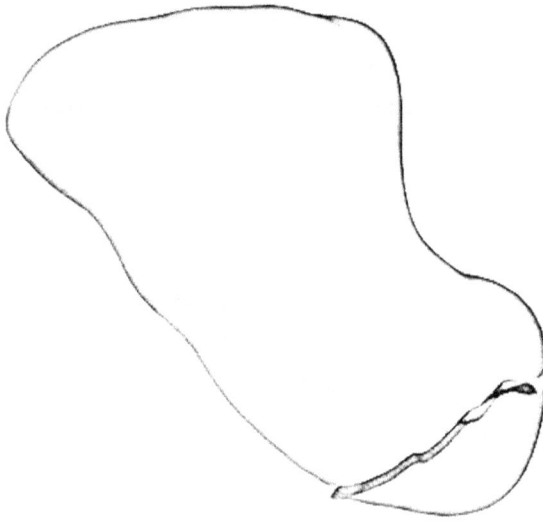

FIGURE 5.5   Proximal fracture in a scaphoid bone.

FIGURE 5.6   K-wire.

are also used by surgeons as devices for temporary fixations, drilling, and the insertion of other implants.

5.2.2 HEADLESS SCREWS

A headless screw [6] is a device often used for the internal fixation of scaphoid fractures. It is a headless screw with threads present only on the ends. The threads of the two ends have a different pitch, which causes compression of the fractured bone fragments when the screw is tightened.

Rendered images of a headless screw are shown in Figures 5.7 and 5.8. The screw is designed to have a hollow body and a hexagonal pattern slot in its head. A guidewire is inserted into the bone before fixing the headless screw; this guidewire fits inside the hollow body of the screw to provide a defined line for drilling the screw. The hexagonal slot on the screw's head grips the screw driver during the screw fixation.

5.3 FINITE ELEMENT ANALYSIS

5.3.1 FEA

Finite element analysis is the methodology of developing computational models to simulate and understand how an object might behave under various physical conditions.

FEA is an ideal choice to study the feasibility of medical implants before their production and use in treatment. FEA can help determine the force at which a designed implant fails and if it is not as per the requirement, the implant can be redesigned.

The following matrix equation [7] is solved in FEA using the computational power of commercial software packages such as ANYS:

FIGURE 5.7 Headless screw (top).

FIGURE 5.8 Headless screw (bottom).

$$\{F\} = [K] \cdot \{U\} \qquad (5.1)$$

where $\{F\}$ is the known loads;

  $[K]$ is the known geometry, material properties, and elements; and
  $\{U\}$ is the displacement (to be calculated by solving the equation).

  This equation is mathematically solved using the matrix inversion method:

$$\{U\} = [K]^{-1} \cdot \{F\} \qquad (5.2)$$

In this chapter, we have used the commercial software package ANSYS Student 2023 R1 for an FEA of K-wire and a headless screw, two popular medical implants used in surgery for scaphoid bone fractures.

5.3.2 K-WIRE

5.3.2.1 Geometry

A computer-aided design (CAD) model of a K-wire can be built in SOLIDWORKS by obtaining the precise measurements of a K-wire. The exact parameters differ on the basis of the diameter and the length of the K-wire.

  To design a K-wire, a circle of diameter 1.10 mm inscribing a triangle was sketched using the *circle and line tool*, as shown in Figure 5.9. Subsequently, the *extrude tool* was used to extrude the triangle up to a height of 15 cm, as shown in Figure 5.10.

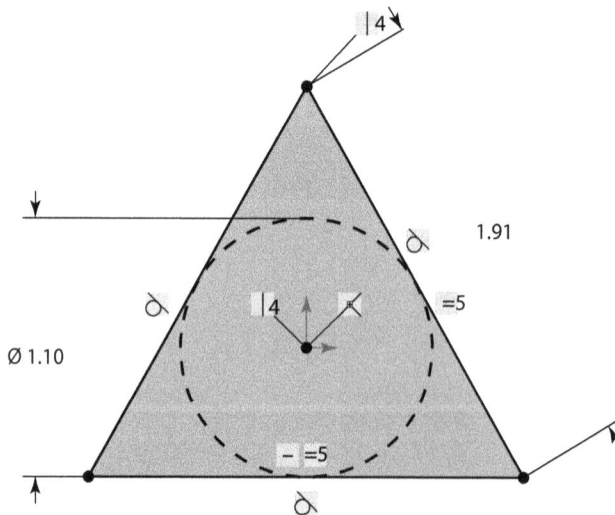

FIGURE 5.9 Dimensions for CAD of a base sketch in a K-wire.

FIGURE 5.10 Extrude tool for CAD of a K-wire.

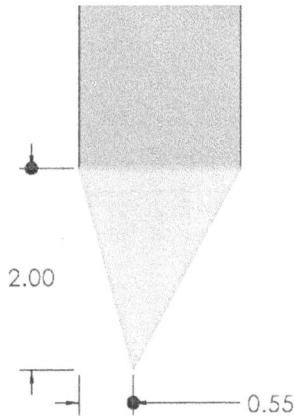

FIGURE 5.11 Dimensions for CAD of chamfer in a K-wire.

The *chamfer tool* was used to design the two trocar-shaped ends of the K-wire with the triangle dimensions, as shown in Figure 5.11. Next, the *fillet tool* was used to round off the edges and complete the CAD model of the K-wire, as shown in Figure 5.12.

FIGURE 5.12 CAD model of a K-wire.

TABLE 5.1
Material Properties of a K-Wire

Implant	Density (kg/m³)	Young's modulus (GPa)	Poisson's ratio	Yield strength (MPa)
K-wire	8000	193	0.31	170

5.3.2.2   Material Properties

The K-wires used by orthopedic surgeons are made of stainless steel [8] and the corresponding material properties used in FEA simulations are shown in Table 5.1.

5.3.2.3   Mesh

For the headless screw, a fine mesh of size 0.2 mm with 40,159 nodes and 20,768 elements was constructed.

5.3.2.4   Loads and Boundary Conditions

The implants used in scaphoid bone surgery need to survive in tension and shear. However, as the role of the K-wire is to restrict the relative motion between the fractured bone parts in the lateral plane, it is necessary that the K-wire survives lateral forces.

Loads were applied on the top cut faces of the K-wire and the bottom cut faces of the K-wire were fixed in 3D space, as shown in Figures 5.13–5.15. The top faces were selected for loading and boundary conditions to represent the worst-case scenario.

5.3.2.4.1   Longitudinal Forces
The K-wire was tested with a longitudinal force varying from 50 N to 85 N to find the best approximation of the force at which the K-wire fails in tension. Figure 5.14 shows the longitudinal force applied on the K-wire.

5.3.2.4.2   Lateral Forces
The K-wire was tested with lateral forces varying from 0.05 N to 0.15 N to find the best approximation of the force at which the K-wire fails in shear. Figure 5.15 shows the lateral force applied on the K-wire.

5.3.2.5   Results
5.3.2.5.1   Longitudinal Forces
The FEA results of total deformation (TDef) and factor of safety (FOS) in a longitudinal force test case of a K-wire are shown in Figures 5.16 and 5.17, respectively. The results showcased in the figures represent the extreme case before failure of the implant.

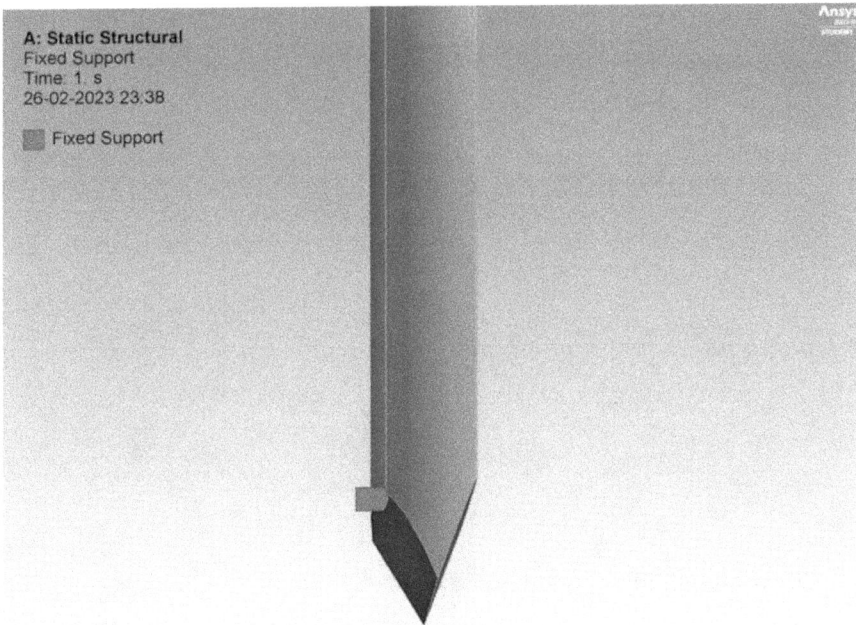

FIGURE 5.13   Fixed support of a K-wire.

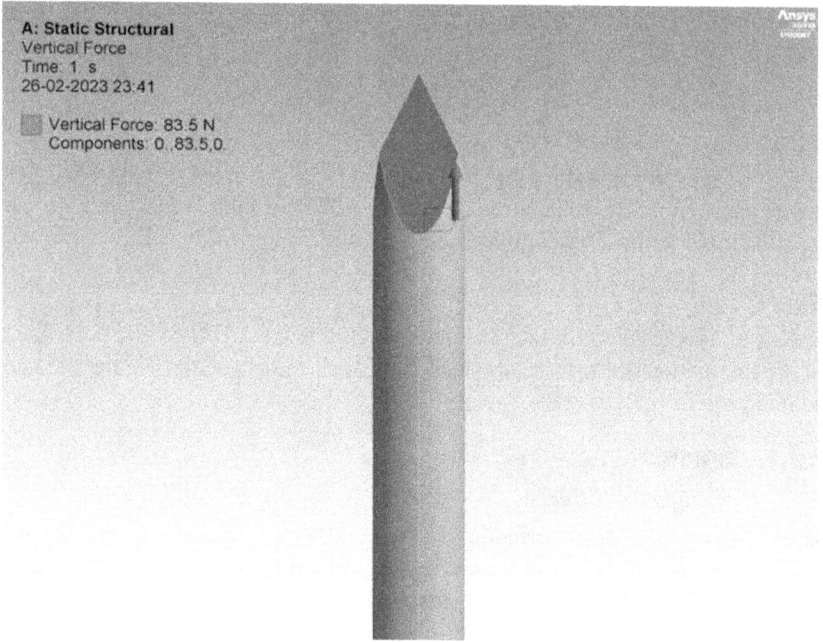

FIGURE 5.14 Longitudinal force on a K-wire.

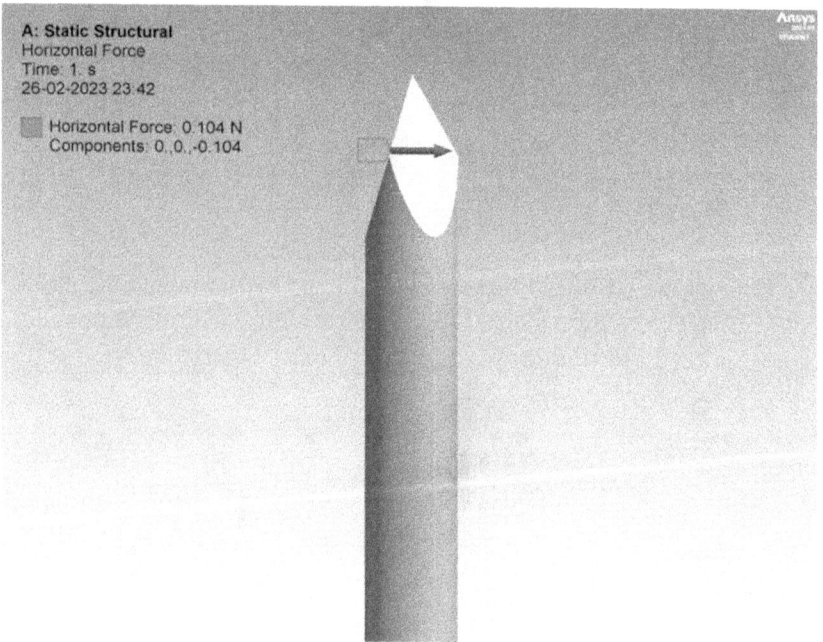

FIGURE 5.15 Lateral force on a K-wire.

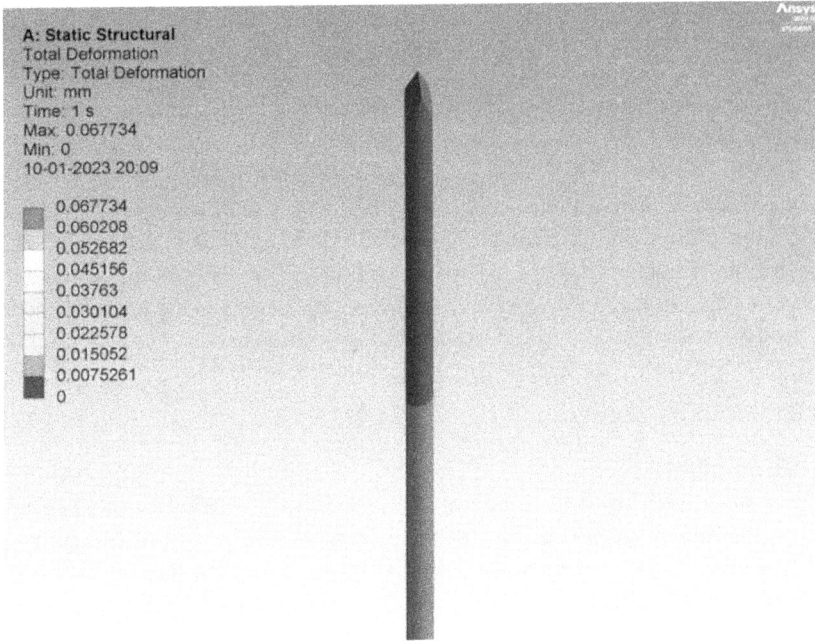

FIGURE 5.16   Deformation in a K-wire at 83.5 N longitudinal force.

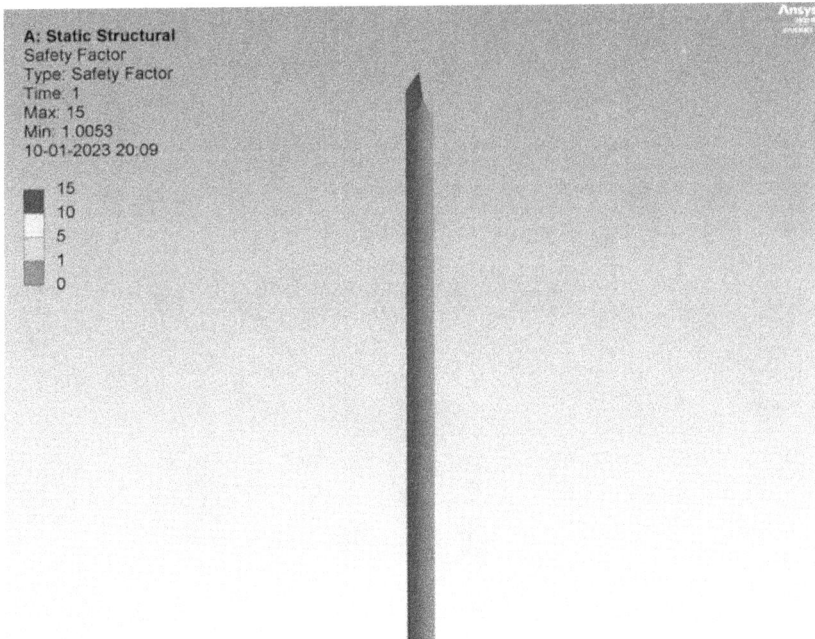

FIGURE 5.17   Factor of safety in a K-wire at 83.5 N longitudinal force.

At a longitudinal load of 83.5 N, the K-wire reached a FOS of 1.0053 and had a deformation of 0.6773 mm. The K-wire deformation in this case is in the longitudinal direction, i.e., increase in length.

5.3.2.5.2 Lateral Forces

The FEA results of total deformation and factor of safety in a lateral force test case of a K-wire are shown in Figures 5.18 and 5.19, respectively. The results showcased in the figures represent the extreme case before failure of the implant.

At a lateral load of 0.104 N, the K-wire reached a FOS of 1.0031 and had a deformation of 8.044 mm. The K-wire deformation in this case is in the lateral direction, i.e., the displacement of its top end from its initial position in the horizontal direction.

5.3.3 HEADLESS SCREWS

5.3.3.1 Geometry

A CAD model of a headless screw was built in SOLIDWORKS by taking precise measurements of a headless screw implant collected from orthopedic surgeons.

A hollow cylinder of height 28 mm was designed with an internal radius of 1 mm and an external radius of 1.45 mm using the *line and extrude tool*, as shown in Figures 5.20 and 5.21. Subsequently, the upper thread was designed with a pitch of 1 mm (Figure 5.22) and the lower thread was designed with a pitch of 1.5 mm

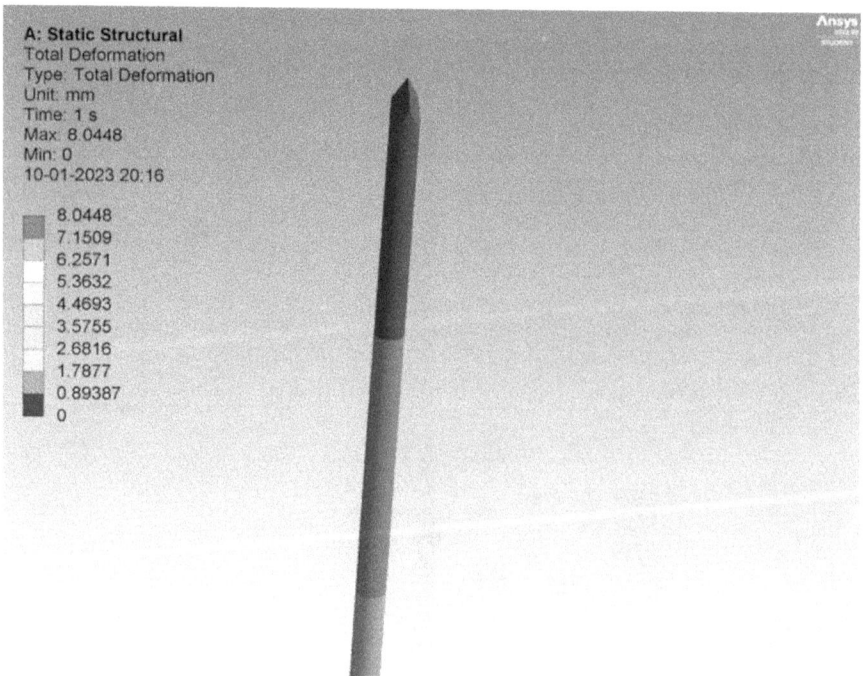

FIGURE 5.18 Deformation in a K-wire at 0.104 N lateral force.

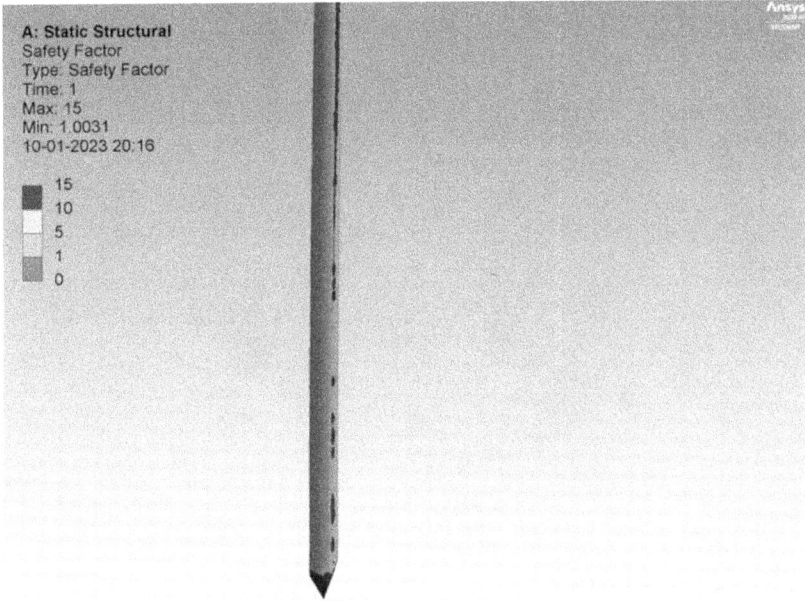

FIGURE 5.19 Factor of safety in a K-wire at 0.104 N lateral force.

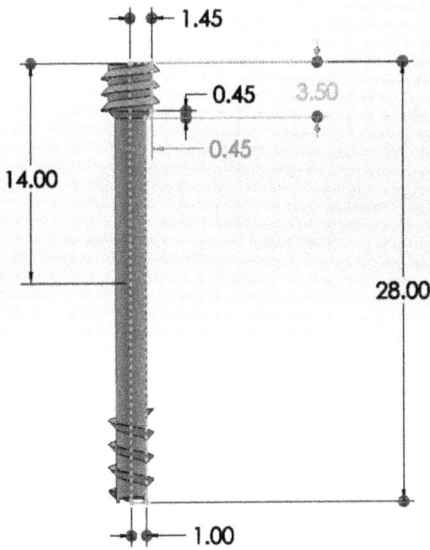

FIGURE 5.20 Dimensions for CAD of a headless screw.

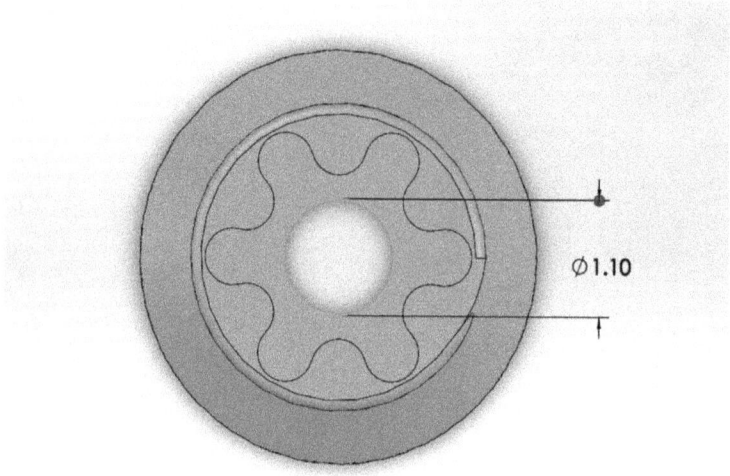

FIGURE 5.21 Dimensions for CAD of screw's head.

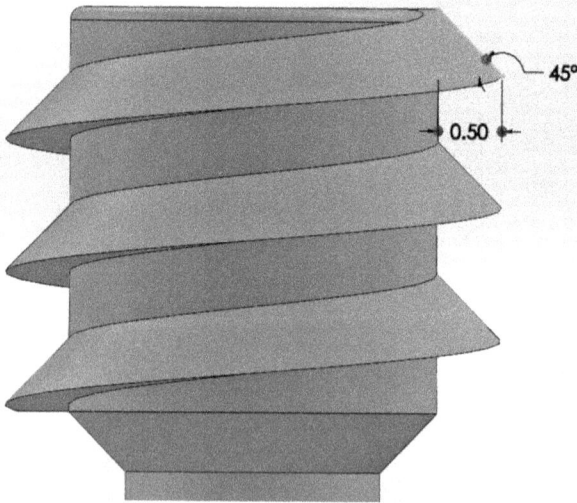

FIGURE 5.22 Dimensions for CAD of the top thread in a headless screw.

(Figure 5.23) using the *helix and spiral* and *swept boss/base tool*. According to the CAD design of a headless screw, with each full rotation of the screw the two fractured bone segments move 0.5 mm closer to one another due to the difference in thread pitch.

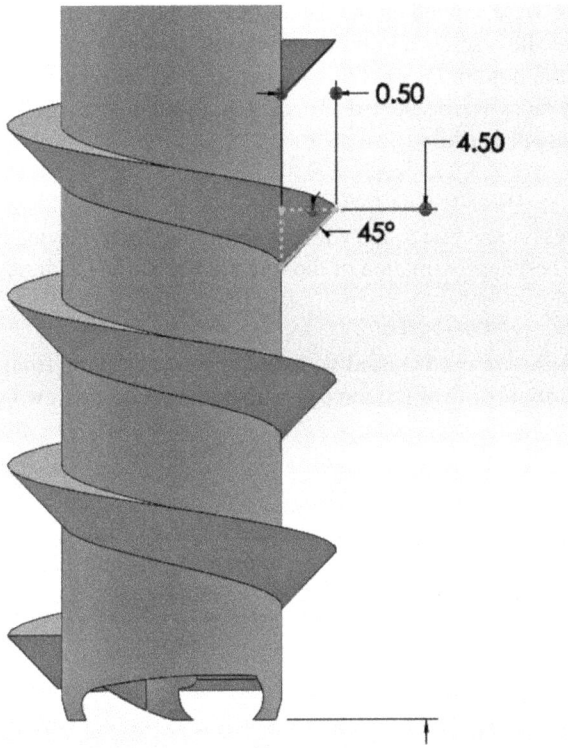

FIGURE 5.23 Dimensions for CAD of the bottom thread in a headless screw.

TABLE 5.2
Material Properties of a Headless Screw

Implant	Density (kg/m³)	Young's modulus (GPa)	Poisson's ratio	Yield strength (MPa)
Headless screw	4500	116	0.34	220

5.3.3.2 Material Properties
The headless screw used by orthopedic surgeons is made of pure titanium [9] and the corresponding material properties used in FEA simulations are shown in Table 5.2.

5.3.3.3 Mesh
A fine mesh of size 0.2 mm with 32,445 nodes and 18,130 elements was constructed for the headless screw.

5.3.3.4 Loads and Boundary Conditions
The implants used in scaphoid bone surgery need to survive in tension and shear. Loads were applied on the top head of the headless screw and the bottom head was fixed in 3D space.

Loads were applied on the flat surface of the top buttress thread and the top head of the headless screw, as shown in Figures 5.24 and 5.25. Additionally, the flat surface of the bottom buttress thread was fixed in 3D space, as shown in Figure 5.26. The top faces were selected for loading and boundary conditions to represent the worst-case scenario.

5.3.3.4.1 Longitudinal Forces

A headless screw was tested with a longitudinal force varying from 200 N to 225 N to find the best approximation of the force at which the headless screw fails in tension.

5.3.3.4.2 Lateral Forces

A headless screw was tested with lateral forces varying from 5 N to 7 N to find the best approximation of the force at which the headless screw fails in shear.

5.3.3.5 Results

5.3.3.5.1 Longitudinal Forces

The FEA results of total deformation and factor of safety in a longitudinal force test case of a headless screw are shown in Figures 5.27 and 5.28, respectively. The results showcased in the figures represent the extreme case before failure of the implant.

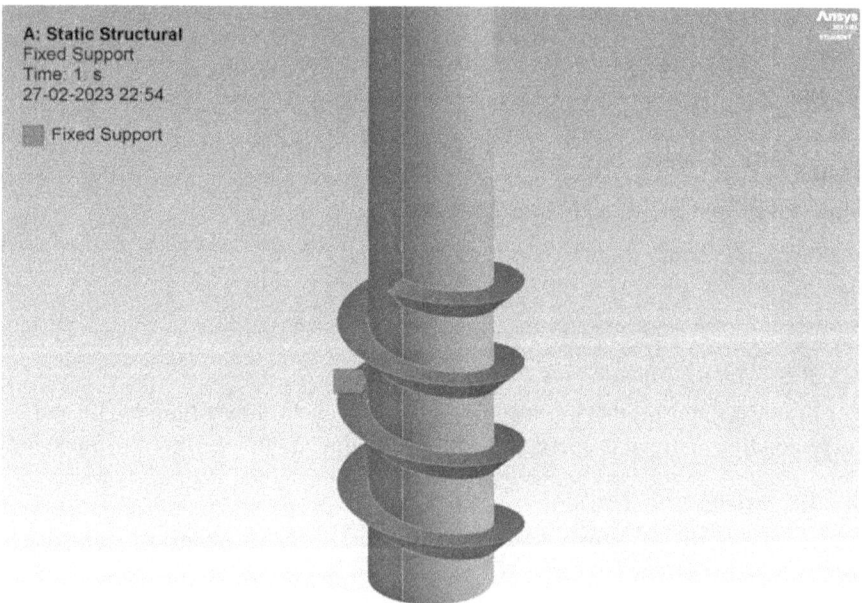

A: Static Structural
Fixed Support
Time: 1. s
27-02-2023 22:54

Fixed Support

FIGURE 5.24 Fixed support of a headless screw.

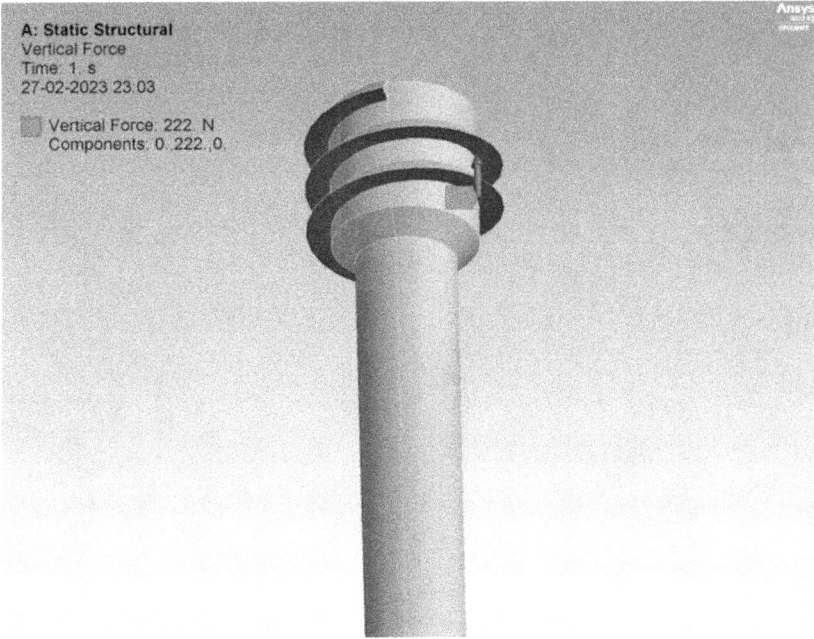

FIGURE 5.25 Longitudinal force on a headless screw.

FIGURE 5.26 Lateral force on a headless screw.

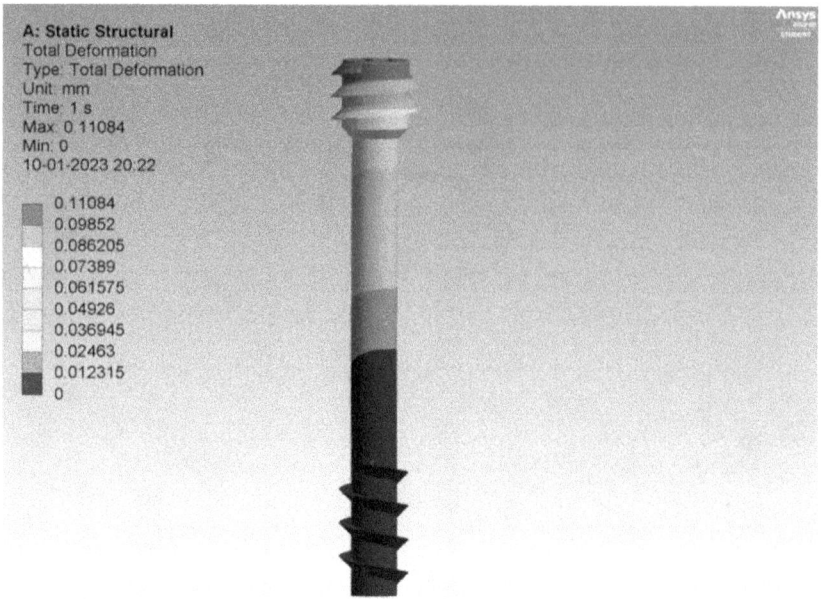

FIGURE 5.27 Deformation in a headless screw at 222 N longitudinal force.

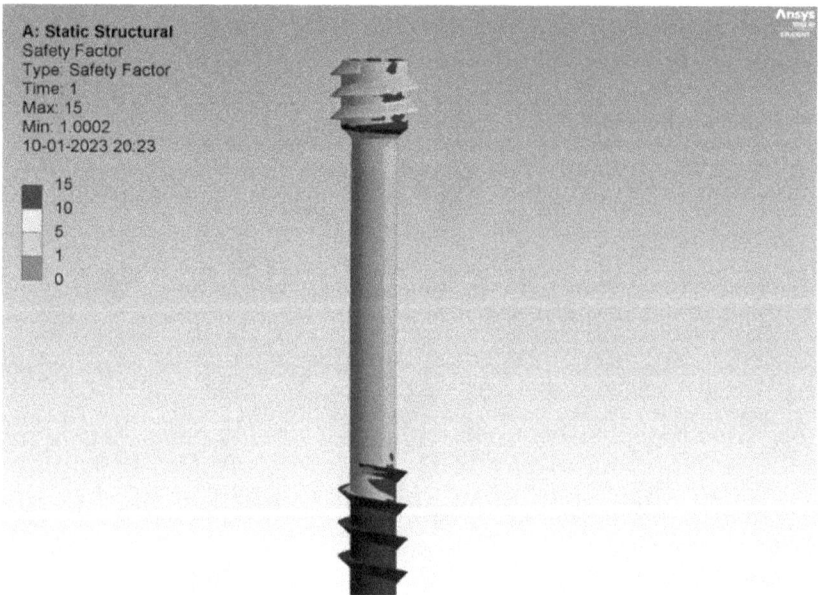

FIGURE 5.28 Factor of safety in a headless screw at 222 N longitudinal force.

At a longitudinal load of 222 N, the headless screw reached a FOS of 1.0002 and had a deformation of 0.11084 mm. The headless screw deformation in this case is in the longitudinal direction, i.e., increase in length.

5.3.3.5.2 Lateral Forces

The FEA results of total deformation and factor of safety in the lateral force test case of a headless screw are shown in Figures 5.29 and 5.30, respectively. The results showcased in the figures represent an extreme case before failure of the implant.

At a lateral load of 6.85 N, the headless screw reached a FOS of 1.0051 and had a deformation of 0.16173 mm. The headless screw deformation in this case is in the lateral direction, i.e., the displacement of its top head from its initial position in the horizontal direction.

5.4 CONCLUSION

The loading of K-wire and headless screw implants leads to strain, which in turn develops stress in these components. The implants start yielding after reaching a stress equal to or greater than the yielding stress of the material, i.e., stainless steel or titanium in this study. In FEA using ANYS, the von Mises stress criteria are used to predict the yielding in materials under complex loading conditions. Thus, total

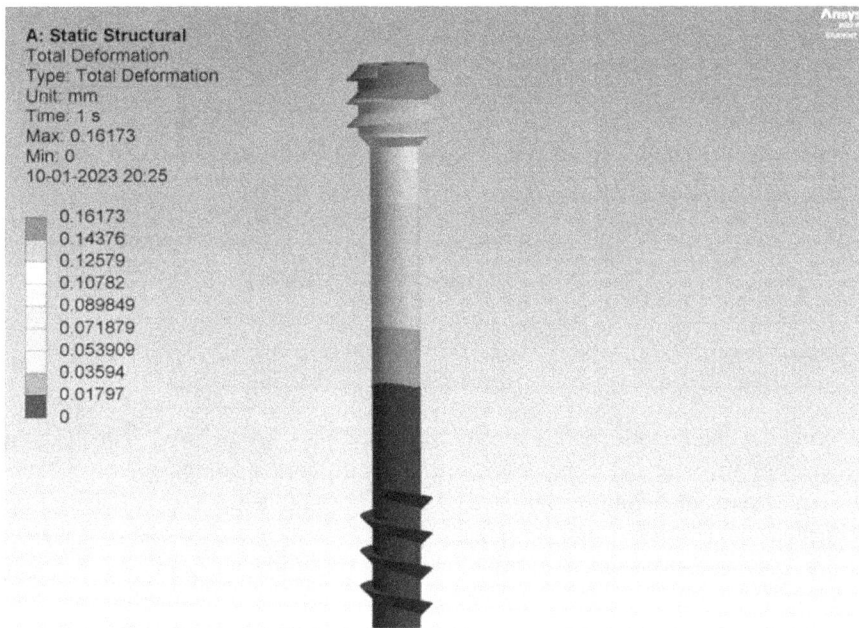

A: Static Structural
Total Deformation
Type: Total Deformation
Unit: mm
Time: 1 s
Max: 0.16173
Min: 0
10-01-2023 20:25

0.16173
0.14376
0.12579
0.10782
0.089849
0.071879
0.053909
0.03594
0.01797
0

FIGURE 5.29 Deformation in a headless screw at 6.85 N lateral force.

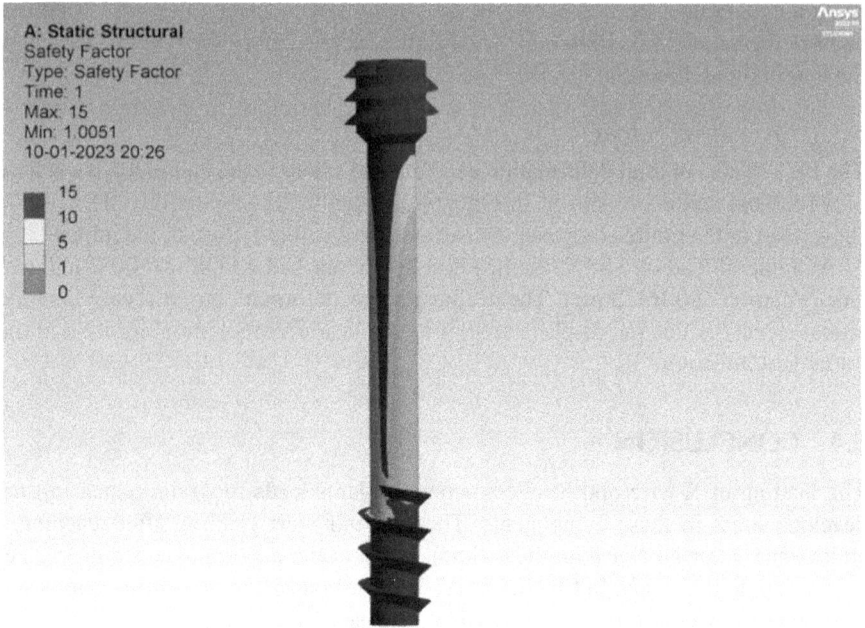

FIGURE 5.30 Factor of safety in a headless screw at 6.85 N lateral force.

TABLE 5.3

Criteria for Failure of a K-Wire and a Headless Screw under Longitudinal and Lateral Force

Test case	K-wire			Headless screw		
	Force (N)	Def (mm)	FOS	Force (N)	Def (mm)	FOS
Longitudinal force	83.50	0.07	1.01	222.00	0.11	1.00
Lateral force	0.10	8.04	1.00	6.85	0.16	1.01

deformation and factor of safety results from ANSYS FEA are used to analyze the behavior of a K-wire and a headless screw under different loading conditions representing extreme real-world scenarios, and a summary of the corresponding results is presented in this section.

The maximum load that K-wire and headless screw implants can tolerate before yielding under tension and shear is summarized in Table 5.3. On the basis of these values it can be conclusively stated that surgeons can use the implant for recuperation surgery of scaphoid non-union fractures.

In this particular case, it is observed that the K-wire and headless screw can easily withstand a load greater than 50 N and 200 N, respectively, in tension. Furthermore, in the case of shear, the K-wire and headless screw can easily withstand a load greater than 0.05 N and 6 N, respectively.

The survival of these implants at these loads qualifies them for use in recuperation surgeries of scaphoid non-union fractures.

It should be noted that the forces mentioned in Table 5.3 for the failure criteria of implants are calculated under the assumption that the force is acting on a solitary implant. However, in reality, implants are inserted inside bones and thus would survive a much higher force in each test case due to the additional strength and support provided by the bones.

REFERENCES

1. Seth, Ajay, et al. "OpenSim: A musculoskeletal modeling and simulation framework for in silico investigations and exchange". *Procedia IUTAM* 2 (2011): 212–232.
2. Fowler, John R., and Hughes, Thomas B. "Scaphoid fractures". *Clinics in Sports Medicine* 34(1) (2015): 37–50.
3. Rhemrev, Steven J., et al. "Current methods of diagnosis and treatment of scaphoid fractures". *International Journal of Emergency Medicine* 4 (2011): 1–8.
4. Santoshi, J. A., et al. "Mechanism of scaphoid waist fracture: Finite element analysis". *Journal of Hand Surgery (European Volume)* 48(5) (2023): 426–434. doi:10.1177/17531934221145516.
5. Herbert, T. J., Fisher, W. E., and Leicester, A. W. "The Herbert bone screw: A ten year perspective". *The Journal of Hand Surgery: British & European Volume* 17(4) (1992): 415–419.
6. Franssen, B. B., et al. "One century of Kirschner wires and Kirschner wire insertion techniques: A historical review". *Acta Orthopaedica Belgica* 76(1) (2010): 1–6.
7. Roylance, David. *Finite Element Analysis*. Massachusetts Institute of Technology, Cambridge, 2001.
8. AZoM. "Stainless steel - Grade 316L - Properties, fabrication and applications (UNS S31603)". AZoM.com, February 18, 2004.
9. AZoM. "Titanium (Ti) - the different properties and applications". AZoM.com, June 6, 2013.

6 Modeling and Prototyping of Anatomical Structures

Boppana V. Chowdary, Nishkal George, and Lakha Mattoo

6.1 INTRODUCTION

In the education of medical professionals, great emphasis is placed on patient anatomy. However, in recent times, traditional methods of preparation such as cadaver dissection (CD) have been criticized for their efficiency and ethics. Artistically rendered models and illustrations have given way to more technological solutions. However, these solutions compromise the tangible nature of models and CD. For the preoperative study of underlying structures, these electronic tools may not be entirely appropriate [1]. This is especially true when patient-specific data in the form of computerized tomography (CT) and magnetic resonance imaging (MRI) images are readily available.

Rapid prototyping (RP), commonly referred to as 3D printing, is the fabrication of three-dimensional physical models directly from a computer-aided design (CAD) [2]. Technological advances in RP have potential ways to assist in improving anatomical education and patient outcomes with higher reliability than cadavers [3]. Recent studies show that 3D-printed anatomical models have been successfully implemented in many use cases [4]. However, due to the technical requirements and resource usage of the RP process, particularly in developing countries, justification for the use of RP technology is still critical [5].

The purpose of this study is to develop a pragmatic methodology for 3D modeling and subsequent prototyping of a patient-specific anatomical model which could be used for downstream medical applications, while identifying key process issues. Further, the proposed study facilitates a reduction in resource consumption thereby reducing manufacturing costs. To achieve these goals, the study identifies three end-user requirements which when applied to the RP process will yield a model of high integrity that also satisfies manufacturing requirements. For this study, CT scans of a human skull with an anomaly are used to generate a parametric CAD model by means of Mimics 3-matic design software. This data is then exported to a fused deposition modeling (FDM) machine where a physical model of the anatomical structure is manufactured.

DOI: 10.1201/9781003375098-6

6.2 LITERATURE REVIEW

In recent years, the traditional use of CD as a surgical planning and educational tool for human anatomy has greatly reduced due to its inefficiency and changing social mores [6]. Alternative forms of anatomical structure representation have thus emerged. Artistic anatomical models and 3D virtual software are quite common alternatives for medical students [7], but these come with critical limitations, such as the surgical navigation of physical structures. Common anatomical features of the human body can be easily understood using these tools, but difficulty arises in the determination of complex cases. Additionally, although an excellent understanding of a patient's anatomy is important, medical trainees should also have knowledge of the spatial relationships to surrounding structures such as the underlying neural tissue framework [1].

Despite promising case studies available in the literature, conventional 2D and 3D visualization tools are still limited to a 2D screen, which restricts a true representation of anatomical structures, and this is particularly apparent in the training of medical students. This has created tremendous avenues for the application of RP/3D printing principles in the medical field [8,9]. With the arrival of new 3D printing technology, 3D-printed anatomical models have been successfully utilized in surgical training, preoperative planning, surgical simulation, diagnosis, and treatments [4]. The typical subtractive manufacturing process involves the shaping of a block of raw material, which lacks the ability to accurately recreate the intricate detailing of internal anatomical features [10]. In general, the RP process mitigates numerous structural design issues that arise in traditional subtractive processes [11].

A key element of the proposed research is capturing the patient's CT or MRI scans that will be consistently utilized in tandem with the RP process. In endovascular surgeries, the anatomical detail derived from patient-specific CT data has proved to be more reliable for preoperative planning than a cadaver model [3]. Studies researching other additive manufacturing (AM) techniques have also used CT scans. For instance, data acquired from CT was reverse engineered to generate a CAD model that was fabricated using a JS-PolyJet machine [12]. A suitable framework for capturing and utilizing CT data to generate patient-specific models was refined to model a temporal bone for instructional and preoperative planning [5].

The accuracy of these models has also been described in the literature. A recent review conducted by Sun and Liu [8] found that 3D-printed kidney models are very precise in describing renal anatomical structures and renal tumors with high precision. Further, orthopedic practitioners are actively using the RP technique to design patient-specific models, instrumentation, implants, orthosis, and prosthesis [13]. Despite the current success, there is still a need to conduct further research on resource-efficient modeling and the prototyping of anatomical structures that provide an accurate transfer of knowledge.

The methods described in these studies demonstrate the usefulness of 3D printing in the medical space and give relative assurance that this technology is low risk and high reward. However, the gap observed is the need to provide empirical evidence supporting the transition to this technology for developing countries [5]. The literature also shows the economic advantages of the RP process combined with CT data for manufacturing anatomical models. This gap may be improved by reconciling end-user requirements with resource efficiency in terms of material and time.

6.3 MATERIALS AND METHODS

The proposed methodology for the development of a CT-derived RP model is shown in Figure 6.1. Twelve steps are proposed to achieve an optimal patient-specific CT model. Steps (i)–(v) are categorized as "setup" tasks; steps (vi) and (vii) are hollowing and inspection operations; steps (viii) and (ix) are the finishing operations; and steps (x)–(xii) are the execution of the RP model development. Further, these steps provide a detailed account of the definitions, tools, and techniques employed by researchers. Additionally, the approach followed in the execution of this research greatly depends on validation by means of established criteria that are well documented in

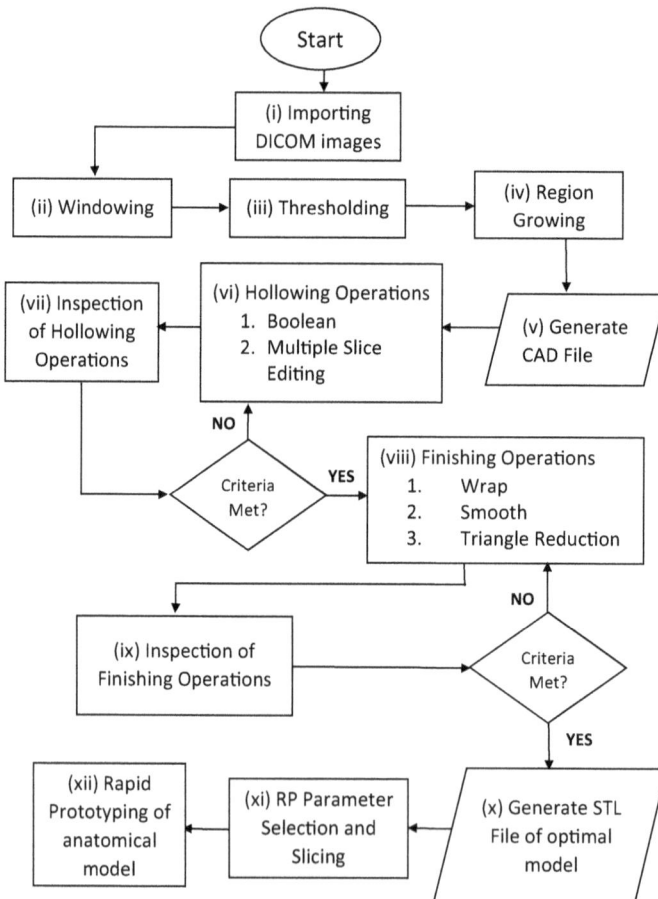

FIGURE 6.1 Methodology for the development of a CT-derived anatomical prototype.

the literature supported by user decisions throughout the process. A detailed explanation of these steps is provided in the following sections.

6.3.1 STEPS (I)–(V): SETUP OF CT VISUALIZATION

CT is an incredibly useful imaging method for preoperative planning when combined with image segmentation and CAD tools. The initial setup operations for CT images begin with the importation of patient files, held in standard digital imaging and communications in medicine (DICOM) formats. This task was accomplished by means of the Mimics surgical standard tessellation language (STL) manipulation software [14]. The images are then registered in the design window of the image processing software. This process is referred to as windowing, and also involves CT registration. Thresholding is the process of selecting the required Hounsfield unit (HU) limits which coincide with the density of the tissue being extracted from the CT data. In this study, the HU limits for a CT image of the skull were set as per a previous study [15]. This selection is then extracted from the CT images through region growing, which produces a better skull model by removing floating pixels. Additionally, to enhance the segmentation process, Mimics 3-matic tools such as Edit Mask and Crop Mask were used to threshold the selected skull mask.

6.3.2 STEP (VI): HOLLOWING OPERATIONS

The ability to remove unnecessary data from the segmented images can be useful for improving resource efficiency. In this case, the pixels removed effectively create an empty space between solid walls which is expected to impact the material usage and production time. One hollowing tool, identified in the image processing software as the Multiple Slice Edit tool of Mimics, allows for the selection of an outer shell of the skull model by presenting a range of pixel values directly related to the thickness of the skull. To refine the developed skull model, the number of pixels was varied between 1 and 50 to fulfil the design and manufacturing requirements. A set of generated feasible skull models are shown in Figure 6.2.

The images shown in Figure 6.2 display the impact of varying pixels on the model. The selected region indicated shows the selected mask volume to be removed to hollow the model. The light gray region shows the original skull. As observed in Figure 6.2a, a large amount of material was selected for removal whereas for Figure 6.2d, only a small amount of material was selected for removal. A pixel selection of 30 or greater was found to exceed the model limits and provided no useful data. Figure 6.3 shows the effect of variations in the pixel number on the mask volume. From the graph it can be observed that a significant reduction in mask volume was accomplished with the least pixel value selection of 1–5.

At this stage of the study, for better visualization of the selected skull masks, the Boolean tool of Mimics program was applied. Further, the Boolean tool supports the morphology operation that produces a shell of desired thickness. In fact, the Boolean tool removes the internal pixels if the thickness of the generated mask exceeds the

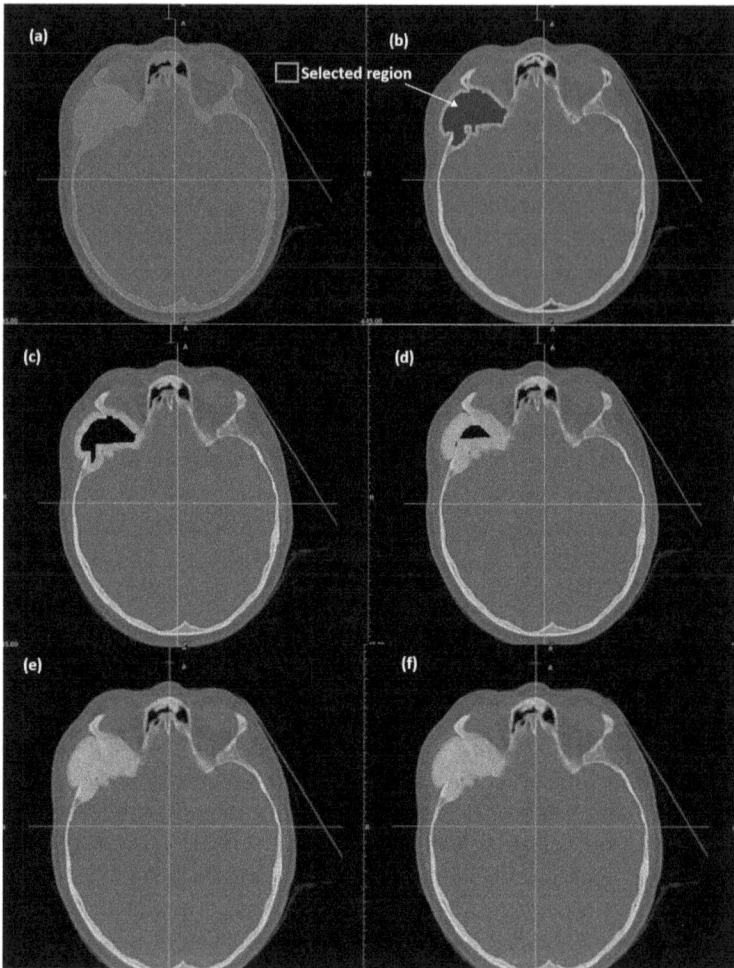

FIGURE 6.2 Cross section of the original skull model after execution of the morphology operation with variation in pixel values: (a) 1 pixel, (b) 5 pixel, (c) 10 pixel, (d) 20 pixel, (e) 30 pixel, and (f) no pixel selection.

user-specified thickness. Figure 6.4 shows various cross sections of the skull model after execution of the Boolean operation with pixel values of 0, 1, 5, 10, 20, and 30. The gray region represents the original segmented area, and the selected region is the remaining area after the Boolean operation.

The morphology operation selects the region to be removed while the Boolean operation removes the selected area. It is important to note that Multiple Slice Edit and Boolean operations are in 2D and hence error detection is difficult. Therefore, visual inspection was performed in the study after the initial 3D model had been rendered (Figure 6.5).

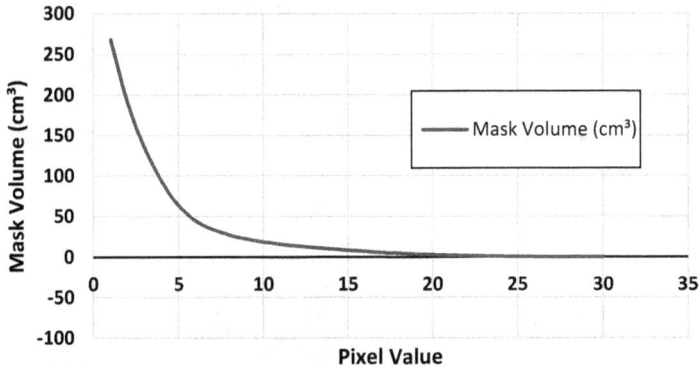

FIGURE 6.3 The impact of variations in pixels on the mask volume for the morphology operation.

6.3.3 STEP (VII): VISUAL INSPECTION OF HOLLOWING OPERATIONS

Through market research, the study determined the most desired end-user needs for an RP anatomical skull model for medical instruction. These needs were then used to develop metrics. A summary of the end-user needs and metrics for the anatomical model, in a typical user need versus metric format is shown in Table 6.1.

Figure 6.6 shows a sample inspection process performed to assess the generated 3D images to ensure each generated model conforms to fulfil the end-user requirements as listed in Table 6.1. The initial step was to generate each model and render the CT data. For example, in Figure 6.6, the highlighted region shows the section where the model was visually analyzed for defects in the form of holes. A count of the number of holes was recorded. Due to the resolution of the images, the defects, i.e., number of holes in the section chosen, were an approximated value. Additionally, the highlighted section in Figure 6.6 shows the thin wall region where criteria 3a and 3b involving model hollowness were assessed. The outcome of the model performance after deployment of the hollowing operation can be seen in Table 6.2.

Defects in CAD models can be expected in various forms, typically in the form of either gaps or holes. As expected, from the data shown in Table 6.2, it was observed that with a 1-pixel value, the developed model exhibits many defects in the form of holes. Further, it is clear that when the pixel value increases the model features tend toward the original behavior of the non-hollowed model. Thus, the authors took the decision to develop a skull model that meets the end-user requirements of being hollow yet maintaining structural integrity with the highest accuracy possible.

Further, it was observed that even as the pixel limit of 30 was approached, some defects or holes, approximately nine, were still present. These holes are attributed to the CT data and considered the typical anatomy of the skull, hence they are very much desired. It can also be seen that the models that meet a minimum of two manufacturing inspection criteria are models A–L with pixel values of 20–29, respectively. Since most of the criteria at this stage are measured via visual aid, for a more

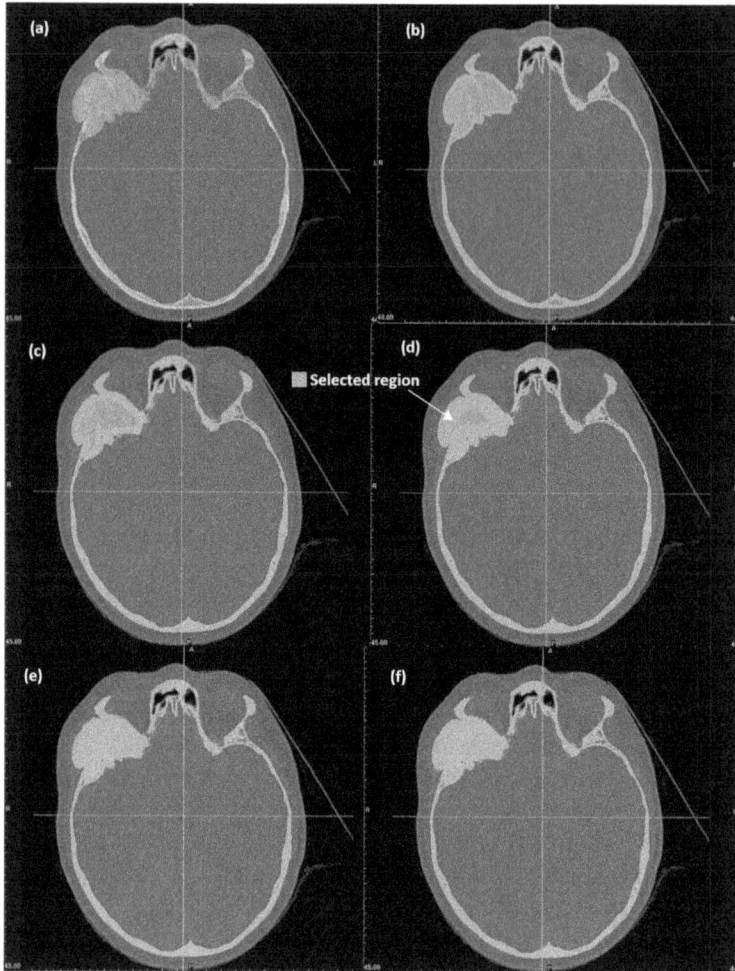

FIGURE 6.4 Various cross sections of the skull model after execution of the Boolean operation with (a) 1 pixel, (b) 5 pixels, (c) 10 pixels, (d) 20 pixels, (e) 30 pixels, and (f) no pixel selection.

quantitative approach to model selection, a volume reduction in the models is now taken into consideration. The main intention of the hollowing process is to cut cost and production time for manufacturing purposes. As per Ulrich and Eppinger [16], a concept scoring matrix was developed for the evaluation of refined skull models. Table 6.3 shows the concept scoring matrix after performing the 3-matic hollowing operation. Model A with a pixel value of 19 is also examined as a benchmark to compare the selected models as it is non-compliant with user needs. To complete the finishing operations step of the design process, as shown in Table 6.3, only models B–E will be used since these scored highest based on the selection criteria.

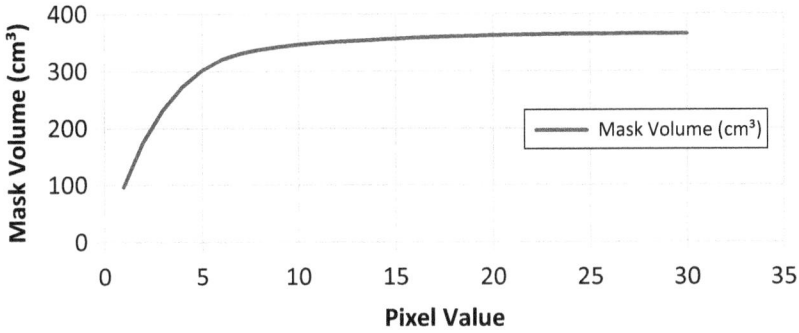

FIGURE 6.5 Mask volume of the shell as it varies with pixel selection.

TABLE 6.1
End-User Needs and Metrics for the Development of Educational Anatomical Skull Models

No.	End-user need	Basis for assessment of end-user need	Metric	Unit
1	Must resemble obtained CT anatomy	CAD image	No. of holes in the selected region	Count
2	Must be accurate and detailed	CAD image		
3	Model must have reduced material volume	(a) CAD image	(a) Hollowness of selected wall region	Subjective
		(b) RP model	(b) Model volume reduction	Cubic meters

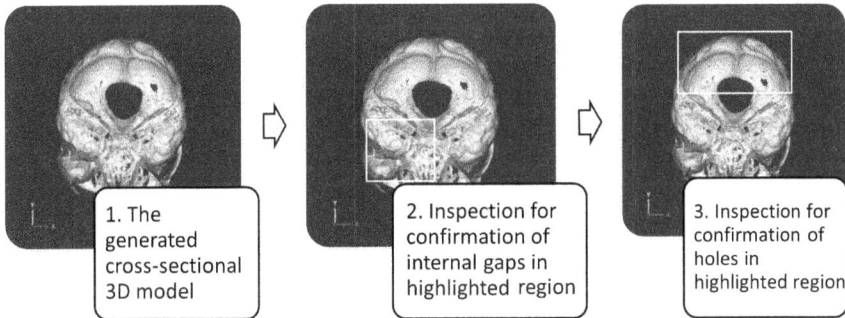

FIGURE 6.6 Sample inspection performed for the fulfilment of user needs 1 and 2 (as per Table 6.1).

TABLE 6.2

Model Performance after Deployment of Hollowing Operation

Pixel value	Volume (cm³)	Surface area (cm²)	Percentage volume reduction	No. of holes in inspected region	Study criteria 1	2	3
1	106.76	4259.62	70.74	194	x	x	√
2	188.35	3468.49	48.37	146	x	x	√
3	242.61	3082.20	33.50	52	x	x	√
4	280.85	2794.68	23.02	34	x	x	√
5	306.67	2557.64	15.94	30	x	x	√
6	322.08	2357.20	11.72	29	x	x	√
7	331.03	2254.15	9.26	26	x	x	√
8	336.92	2197.14	7.65	25	x	x	√
9	341.14	2162.92	6.49	21	x	x	√
10	344.28	2137.74	5.63	18	x	x	√
11	346.75	2125.32	4.95	15	x	x	√
12	348.89	2118.00	4.37	15	x	x	√
13	350.78	2111.69	3.85	14	x	x	√
14	352.42	2105.72	3.40	14	x	x	√
15	353.87	2101.60	3.00	14	x	x	√
16	355.16	2098.18	2.65	12	x	x	√
17	356.33	2095.19	2.33	12	x	x	√
18	357.37	2092.52	2.04	12	x	x	√
19	358.30	2095.52	1.79	10	x	x	√
20	359.10	2086.33	1.57	9	√	√	√
21	359.75	2083.17	1.39	9	√	√	√
22	360.28	2080.27	1.24	9	√	√	√
23	360.70	2077.35	1.13	9	√	√	√
24	361.02	2075.48	1.04	9	√	√	√
25	361.27	2073.88	0.97	9	√	√	x
26	361.46	2072.25	0.92	9	√	√	x
27	361.58	2070.89	0.89	9	√	√	x
28	361.64	2069.80	0.87	9	√	√	x
29	361.37	2069.08	0.95	9	√	√	x
30	361.68	2068.83	0.86	9	√	√	x

√: Model meets criteria as per Table 6.1.
x: Model fails to meet criteria as per Table 6.1.

6.3.4 STEP (VIII): MODEL FINISHING OPERATIONS

Model finishing is the process of tailoring the surface of a model to meet the end-user needs. In this regard, various medical CAD finishing operations such as smoothing, wrapping, and triangle reduction were applied on the developed model. In addition,

TABLE 6.3

Concept Scoring Matrix after Performing a Hollowing Operation

						Concept						
	Model	A	B	C	D	E	G	H	I	J	K	L
Selection criteria	Pixel	19	20	21	22	23	24	25	26	27	28	29
(1) Must resemble obtained CT anatomy	Rating (max 5)	0	4	4	4	4	3	3	3	3	3	3
(2) Must be accurate and detailed		0	4	4	4	4	4	3	3	3	3	3
(3) Model must have reduced volume		4	3	3	2	2	1	0	0	0	0	0
Total score rank		4	11	11	10	10	8	6	6	6	6	6
Develop?		Yes[a]	Yes	Yes	Yes	Yes	No	No	No	No	No	No

[a] Selected as benchmark.

TABLE 6.4

Finishing Operations Values

Operation	Value
Triangular reduction	
• (i) Tolerance (mm)	0.03
• (ii) Edge angle (°)	15.00
Wrap	
• (i) Smallest detail (mm)	0.5
• (ii) Gap closing distance (mm)	0.5
Smoothing factor	0.4

the default settings of these operations, as given in Table 6.4, were used to perform the experiments. Further, five models (A–E) were randomly selected to test the three selected finishing operations and the effect of these operations on model integrity were investigated. Furthermore, the original model was compared with the refined models, and their performance is shown in Figure 6.7.

From Figure 6.7, it can be seen that the finishing operation had the least effect on the volume of the model due to the implementation of the triangle reduction operation. Furthermore, the triangle reduction process improves the mesh quality, thereby reducing the size of the standard tessellation language file, as captured in the study by Zhang et al. [17]. The human skull has a large amount of detail with very small, yet complex bone structures and tiny holes in its anatomy for soft tissues. The smoothing operation effectively levelled out and, in some cases, removed the very

Model Volume of selected models

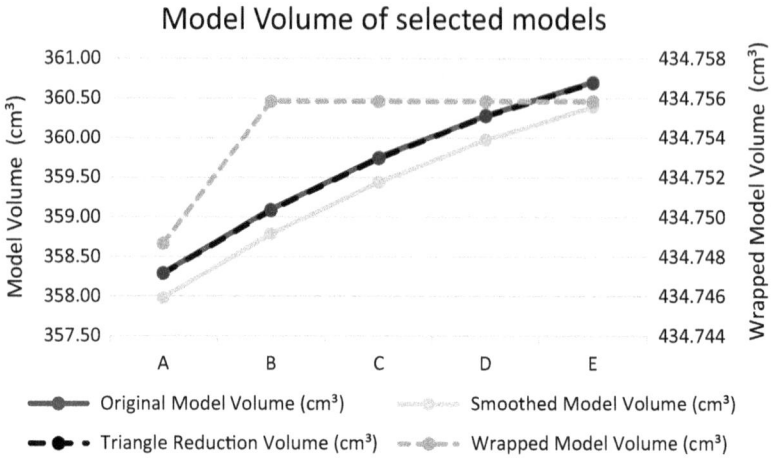

FIGURE 6.7 Comparison of model volume after deployment of finishing operations.

Surface Area of developed models

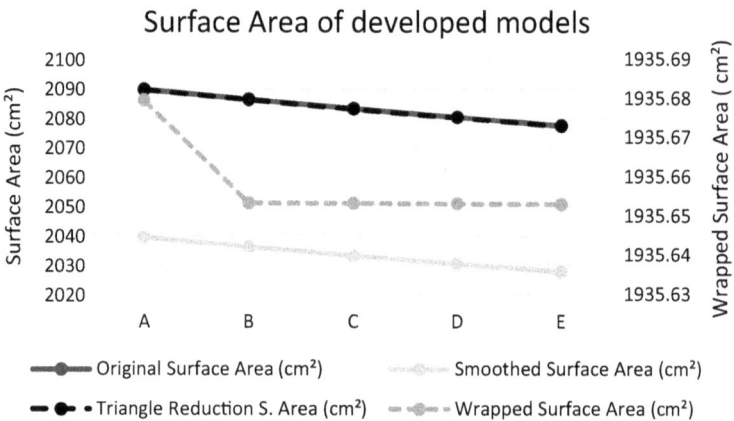

FIGURE 6.8 Comparison of surface area for developed models.

detailed aspects of skull CT data. This could be observed by the significant decrease in the models' volume after performing smoothing operations. The wrapped data, presented on a secondary axis, showed a steady rise in model volume peaking after the 20-pixel value, and subsequently levelling out. In all cases, the wrapping operation increased the model volume. In conclusion, the skull CT data contains many naturally occurring holes as part of its anatomy which could be lost by performing an excessive wrapping operation.

A comparison of the surface area for developed models A–E can be seen in Figure 6.8, which shows that the triangle reduction operation made an insignificant change on the model surface area, while models subjected to smoothing exhibited a significant decrease in surface area. As deduced previously, by removing the small detailing of the skull's anatomy using the smoothing operation, the surface area drastically decreased. The smoothing data had to be put on the secondary axis to allow for a comparison. As per the wrapping data, more holes and gaps did not shown up. The remarkable feature in both graphs was the significant reduction in volume and surface area. To compare the integrity of the thin-walled structures, a visual inspection was performed. Figure 6.9 shows cross-sectional images of the skull model after performing the three individual finishing operations as well as the original model. It is clear that both smoothing and triangle reduction model images closely resemble the original model. However, in the case of the wrapped operation model (Figure 6.9c), it was observed that the hollow section of the original skull (Figure 6.9d), shown by the highlighted region, was completely covered over. For this reason, the wrapping operation was not considered a potential finishing operation to develop anatomical models.

From the data gathered through the application of three finishing operations, the triangle reduction process was able to improve the model by reducing holes with minimal impact on the material volume and surface area. Therefore, the triangle

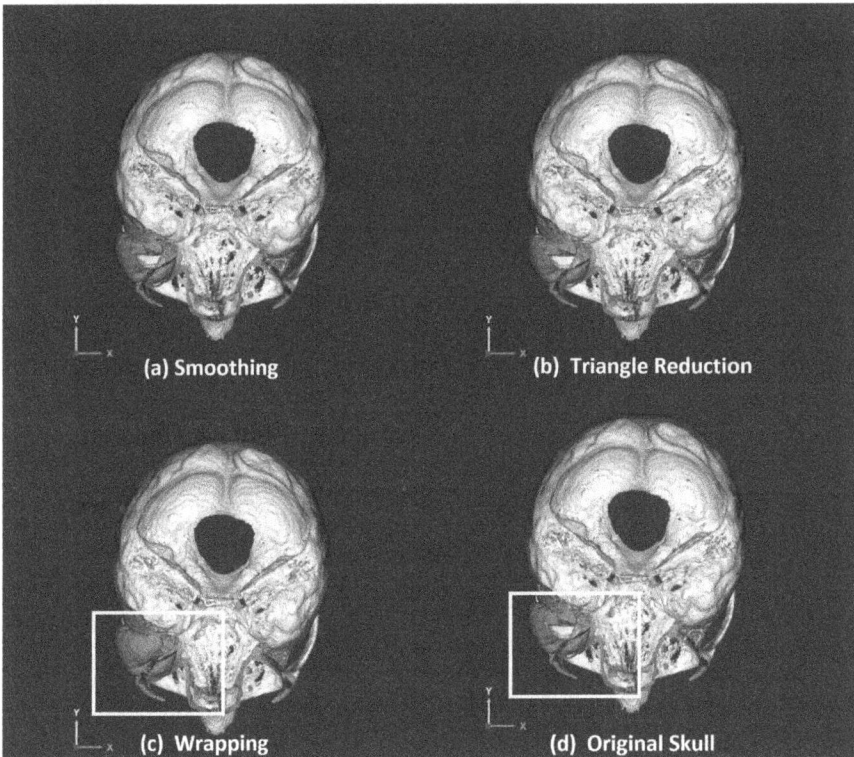

FIGURE 6.9 Sectioned top view of a skull for (a) smoothing, (b) triangle reduction, (c) wrapping, and (d) original model.

reduction tool was chosen as the best finishing operation and, accordingly, the five selected models A–E were further analyzed in the next step.

6.3.5 STEP (IX): INSPECTION OF FINISHING OPERATIONS

In this step, the structure of the finished models is evaluated using the end-user criteria 1 and 2 in Table 6.1. This final inspection ensures that the selected finishing operations performed accurately and maintained the integrity of the external model features.

Figure 6.10 shows models A and B after performing the finishing operations. Model A has a large observable gap that does not pass the inspection process. Table 6.5 shows the results for the five selected models after performing the triangle reduction operation. For validation purposes, five models including a benchmark

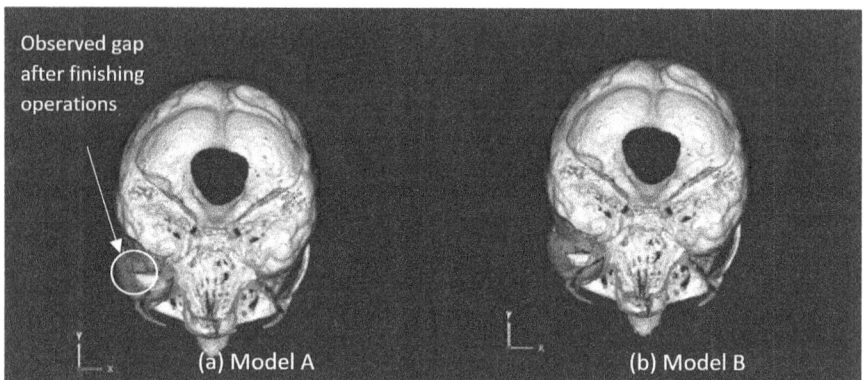

FIGURE 6.10   Comparison of models A and B after performing finishing operations.

TABLE 6.5
Inspection of the Five Selected Models to Validate Finishing Operations

Model	No. of pixels	Volume (cm³)	Surface area (cm²)	No. of holes observed in the model	Criteria met or not?[a]		
					1	2	3
A	19	358.28	2089.46	10	Benchmark		
B	20	359.74	2086.20	9	√	√	√
C	21	360.27	2083.04	9	√	√	√
D	22	360.69	2080.14	9	√	√	√
E	23	360.69	2077.22	9	√	√	√

[a] As per Table 6.1.

TABLE 6.6
Concept Scoring Matrix after Deploying Finishing Operation

		Model				
		Aᵃ	B	C	D	E
Selection criteria	Pixel	19	20	21	22	23
(1) Must resemble obtained CT anatomy	Rating	Benchmark	5	5	4	4
(2) Must be accurate and detailed	(max 5)		5	4	4	4
(3) Model must have reduced volume			4	3	2	2
Total score rank			14	12	10	10
Develop?			Yes	No	No	No

ᵃSelected as benchmark.

were chosen to demonstrate the efficacy of the triangle reduction operation. Table 6.6 shows the concept scoring matrix that was used to identify the optimal model to carry into production Model B was found to be a feasible skull model with the lowest volume that met all three desired criteria.

6.3.6 RAPID PROTOTYPING OF THE SELECTED MODEL

For the development of a prototype, Model B with a pixel value of 20 was selected and converted to STL file format and transferred to the RP workstation. The model was then prepared for printing by slicing, support generation, and tool path creation. The following printing parameters were set in the Stratasys™ Fortus 400mc and the prototype was duly printed with a slice height of 0.01" and T16 nozzle tips for printing the selected anatomical model. Further, the material selected for printing the model was polycarbonate (PC) with PC-10 break-away as the support material. PC was selected as the printing material due to its durability as well as its popularity in several industrial applications such as the medical, construction, and automobile sectors.

The prototype was validated against the Mimics CAD model. Figure 6.11 shows a comparison of different views of the printed skull to the CAD model generated in Materialise Mimics. The printed model exhibits all the intended anatomical characteristics of the CT data. However, in some places of the printed model, particularly in the small crevices of the maxilla and nasal cavities, the supporting material was entrapped in the model. This can be observed in the bottom view of the skull (Figure 6.11d).

Further, the CAD model was measured at specific locations using the Materialise Mimics program and then compared with the corresponding dimensions of the printed model. Figure 6.12 shows the areas selected for measurement and Table 6.7 shows the measured data. It can be observed that the prototype measurements

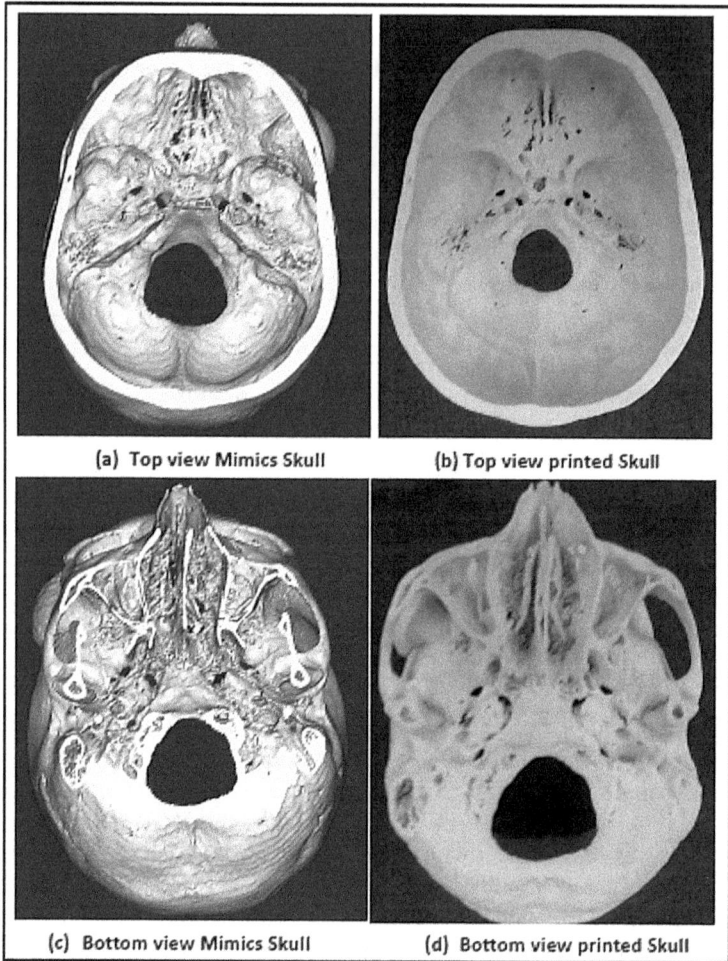

FIGURE 6.11   Comparison of Mimics CAD model with printed prototype.

were similar, suggesting that the model was printed close to the dimensions of the designed anatomical CAD model by deployment of various medical CAD tools and techniques (Table 6.7).

6.4   DISCUSSION

The focus of this research was to develop a methodological approach for the modeling and rapid prototyping of anatomical models from CT data and to deduce possible design issues when developing such models from raw patient-specific data. This was accomplished by fulfilling end-user needs and manufacturing criteria.

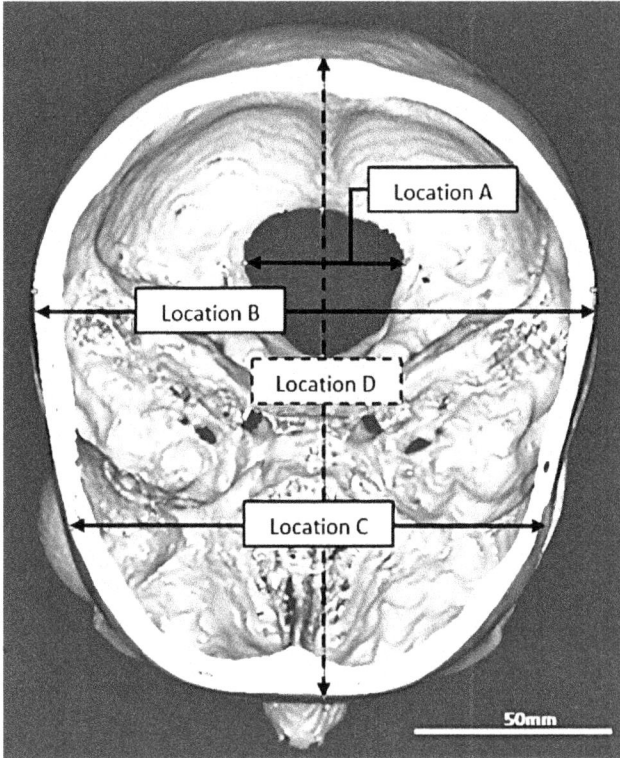

FIGURE 6.12 Measured sections of skull from Materialise Mimics 3D image.

TABLE 6.7
Dimensional Comparison of Mimics CAD Model with the Prototype

Selected location	Measured dimensions (mm)		Error (%)
	Mimics CAD model	Prototype	
A	43.11	43.00	0.26
B	152.45	152.00	0.29
C	127.46	127.00	0.36
D	177.16	177.00	0.09

6.4.1 ISSUES WITH CT DATA PROCESSING

The first issue encountered in processing a patient-specific CT was that of accuracy in the initial stages of data acquisition and registration. Errors were found in the patient's CT scan due to improper calibration of the CT scanner which led to inaccurate results in the downstream modeling operations. This observation concurs with the study of Hajnal and Hill [18]. Other processing issues due to noise, blur, and low contrast were also very common when using CT, leading to inaccurate segmentation of hard and soft tissues. This issue resulted in reduced visibility of small details as described by Al-Ameen and Sulong [19]. Throughout the design phase of the study, these issues were taken into consideration by deploying appropriate medical CAD operations.

6.4.2   Issues with Hollowing Operations

In the morphology phase of the study, the selection of an appropriate number of pixels decides the amount of material to be removed when hollowing the model. Generally, the selected pixel number has a drastic impact on the model anatomy. However, at this stage the model can only be viewed in 2D format, hence defects cannot be seen until it is generated in 3D model form. Therefore, there are no possible solutions to finding an optimal model at this stage of the study. However, through this research an optimal model was found using a trial-and-error method followed by verification of the model structure's integrity.

The execution of the Boolean tool adds another design issue since its operation is dependent on the morphology properties of the model. The Boolean tool removes the selected material which in turn hollows the model. In fact, the developed models at this stage are still in 2D form and cannot provide a concise view of errors that may be present. Automatic hollowing of the model is very difficult since it uses a uniform value to subtract the material. However, an anatomical section of variable thickness, such as the skull in this study, proved difficult when attempting to hollow.

6.4.3   Issues Associated with Model Finishing

While conducting the finishing operations, the preset values were retained to observe the effect on the model. The use of finite values was deemed acceptable for this study, as a range of values would significantly increase the model generation and subsequent analysis time. The triangle reduction method was selected after model prescreening since it did not change the model anatomy while smoothing its surface. Further, it was noted that the wrapping operation could be an ideal choice to improve the model appearance as well as to correct the induced holes due to hollowing.

After execution of the triangle reduction step, a quality inspection was performed to select a feasible anatomical model. From this inspection, Model 20 was chosen as a feasible model since it retained the original anatomical features while being hollowed. But the major problem with this operation is that the skull thickness may be inconsistent and walls that are too thin can cause a weak model.

6.4.4   Issues with Rapid Prototyping of the Anatomical Model

Issues developed when transferring 3D CAD data to RP software. It was noticed that the CAD model gained a small amount of volume (5.02 cm^3) when converted to STL format. This contributes some data inconsistency in the process which cannot be rectified. Additionally, transferring data from RP software to the prototyping machine produces changes in the developed anatomical model structure. This is possibly due to the nature of the selected RP machine and the extruder nozzle size whereby small anatomical features can be lost.

The post-cleaning operation of the model became an issue due to the difficulty of removing the support material. Due to the intricate ridges, small holes, and details, it was difficult to remove all of the support material without damaging the model. As a result, some of the support material was left in the model. This can be mitigated by using a soluble support material and factoring in support location when generating the GCODE. Additionally, a few small gaps were noticed in the prototype. This was a direct result of the printing process where the tool path chosen usually outlines that specific slice of the model and then proceeds to fill it with the support material. Thus, the tiny holes formed were due to the selection of a faulty tool path.

Figure 6.13 shows a comparison of the prototype with the CAD model. From Figure 6.13 it can be noted that the CAD model did not show any holes in the highlighted region. This part of the skull was the thinnest section, which may have contributed to errors since the transferring of the file may have been skewed from the RP software to the prototyping machine. The tool path and the size of the tip used may have also contributed to this error. This can be rectified by choosing a smaller tip size which may increase the chances of these small areas being filled or by choosing a different tool path.

6.4.5 BUILD TIME

The production time is another measure of manufacturability in the RP process and is typically referred to as the build time. Initially, the pixel value was increased incrementally to observe a clear pattern while producing a model. Then, the developed models were simulated in the Insight software to estimate the build time. The build time data, model, and support volumes are shown in Table 6.8.

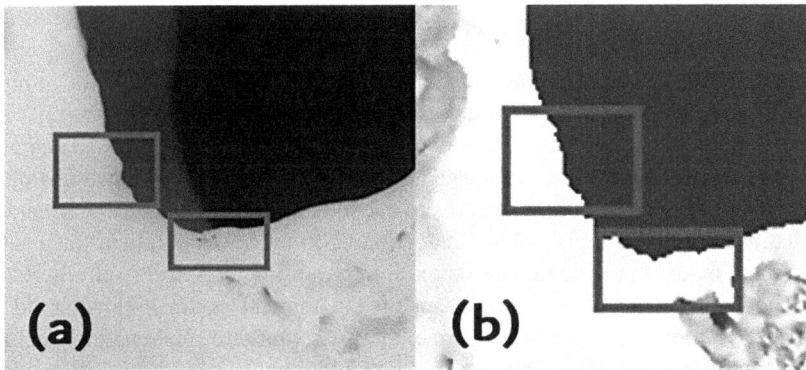

FIGURE 6.13   Printing errors in (a) prototype compared to (b) CAD model.

TABLE 6.8
Build Time Data along with the Estimated Model and Support Volume

Models	Build time (h)	Model volume (cm³)	Support volume (cm³)
1 pixel	49.78	152.99	341.51
5 pixel	33.27	314.35	160.05
10 pixel	29.20	350.47	147.81
15 pixel	28.78	359.40	144.52
20 pixel	28.60	364.12	142.73
25 pixel	28.45	366.12	141.91
Original skull	28.40	366.48	141.91

6.5 CONCLUSION

In this study, a systematic design methodology was proposed to model and prototype a patient-specific anatomical model from CT data to fulfil the end-user and manufacturing criteria. To achieve this, several inspection criteria such as hollowing and triangle reduction were adapted to quantify the impact of parametric CAD operations on the final prototype. The accuracy and build time of the developed models were ranked by means of a concept scoring matrix approach. Various design issues that could negatively impact the proposed methodology were discussed.

By recognizing the design and development issues identified in the study, practitioners can better understand when erroneous events occur. Within a reasonable margin of error, an anatomically accurate RP model was successfully manufactured using a resource-efficient approach. The creation of a patient-specific anatomical model requires a high level of accuracy for its intended purposes. The proposed methodology can enable and encourage the use of RP tools and techniques in the medical sector, particularly to perform preoperative procedures.

6.6 LIMITATIONS AND FUTURE RESEARCH

Though well established, the use of CT data to derive 3D models is not immune from image processing errors. Variations in anatomical structures also present challenges to downstream medical operations. In this study, a CT image–based anatomical RP model was successfully modeled and developed. However, the anatomy of the skull proved difficult to hollow effectively due to its thin-walled nature. Further, the results indicate that a slight volume reduction (2.36 cm³) was obtained for a better anatomical model but a significant increase in build time was observed. Therefore, this study concluded that for a patient-specific anatomical model, hollowing did not significantly impact the production time. However, massive anatomical structures or structures with greater thickness could potentially benefit from the methodology proposed in the study.

Future work should aim at validating the design process through several case studies of bigger anatomical structures and the optimization of manufacturing criteria by means of the established predictive tools and techniques.

REFERENCES

1. Azer SA, Azer S. 3D anatomy models and impact on learning: A review of the quality of the literature. *Health Professions Education*. 2016;2(2):80–98. doi:10.1016/j.hpe.2016.05.002
2. Cooper KG. Rapid Prototyping Technology: Selection and application. *Assembly Automation*. 2001;21(4):358–359. doi:10.1108/aa.2001.21.4.358.1
3. Kaschwich M, Sieren M, Matysiak F, Bouchagiar J, Dell A, Bayer A, et al. Feasibility of an endovascular training and research environment with exchangeable patient specific 3D printed vascular anatomy. *Annals of Anatomy – Anatomischer Anzeiger*. 2020;231:151519. doi:10.1016/j.aanat.2020.151519
4. Yap YL, Tan YS, Tan HK, Peh ZK, Low XY, Yeong WY, et al. 3D printed bio-models for Medical Applications. *Rapid Prototyping Journal*. 2017;23(2):227–235. doi:10.1108/rpj-08-2015-0102
5. Francis V, Ukey P, Nayak A, Taufik M, Jain PK, Mankar SH, et al. Influence of 3D printing technology on biomedical applications: A study on surgical planning, procedures, and training. Lecture Notes in Mechanical. *Engineering*. 2020:269–278. doi:10.1007/978-981-15-4748-5_26
6. Fredieu JR, Kerbo J, Herron M, Klatte R, Cooke M. Anatomical models: A digital revolution. *Medical Science Educator*. 2015;25(2):183–194. doi:10.1007/s40670-015-0115-9
7. Yeom S, Choi-Lundberg DL, Fluck AE, Sale A. Factors influencing undergraduate students' acceptance of a haptic interface for learning gross anatomy. *Interactive Technology and Smart Education*. 2017;14(1):50–66. doi:10.1108/itse-02-2016-0006
8. Sun Z. Insights into 3D printing in medical applications. *Quantitative Imaging in Medicine and Surgery*. 2019 Jan;9(1):1–5. doi:10.21037/qims.2019.01.03
9. Wu AM, Wang K, Wang JS, Chen CH, Yang XD, Ni WF, Hu YZ. The addition of 3D printed models to enhance the teaching and learning of bone spatial anatomy and fractures for undergraduate students: A randomized controlled study. *Annals of Translational Medicine*. 2018 Oct;6(20):430. doi:10.21037/atm.2018.09.59
10. Gibson I, editor. *Advanced Manufacturing Technology for Medical Applications: Reverse Engineering, Software Conversion and Rapid Prototyping*. Chichester: John Wiley & Sons. 2006 Jun 14.
11. Jin Y, Ji S, Li X, Yu J. A scientometric review of hotspots and emerging trends in additive manufacturing. *Journal of Manufacturing Technology Management*. 2017;28(1):18–38. doi:10.1108/jmtm-12-2015-0114
12. Kudasik T, Miechowicz S. A method of reproducing complex, multi-object anatomical structures. *Rapid Prototyping Journal*. 2017;23(1):1–6. doi:10.1108/rpj-05-2015-0058
13. Lal H, Patralekh MK. 3D printing and its applications in orthopaedic trauma: A technological marvel. *Journal of Clinical Orthopaedics and Trauma*. 2018;9(3):260–268. doi:10.1016/j.jcot.2018.07.022
14. Materialise. *Mimics Student Edition Course Book*. Belgium. 2010.
15. Kamalian S, Lev MH, Gupta R. Computed tomography imaging and angiography – Principles. *Handbook of Clinical Neurology*. 2016;135:3–20. doi:10.1016/b978-0-444-53485-9.00001-5
16. Ulrich KT, Eppinger SD, Yang MC. *Product Design and Development*. Boston: McGraw-Hill Higher Education. 2008.

17. Zhang L, Hu Z, Wang MY, Feih S. Hierarchical sheet triply periodic minimal surface lattices: Design, geometric and mechanical performance. Materials & mechanical performance. *Materials and Design.* 2021;209:109931. doi:10.1016/j.matdes.2021.109931

18. Hajnal JV, Hawkes DJ, Hill DLG, editors. *Medical Image Registration.* Boca Raton: CRC Press. 2001.

19. Al-Ameen Z, Sulong G. Prevalent degradations and processing challenges of computed tomography medical images: A compendious analysis. *International Journal of Grid and Distributed Computing.* 2016;9(10):107–118. doi:10.14257/ijgdc.2016.9.10.10

7 Introduction to Conventional and New Generation Manufacturing Processes

R.S. Jadoun and Sushil Kumar Choudhary

7.1 INTRODUCTION

In conventional machining, material is removed via direct contact between the tool and the workpiece. Energy is required to spin both the workpiece and the tool, and the cutting tool is recommended to be 35%–50% harder than the workpiece. Non-traditional machining processes (NTMP) generally utilizes thermal, mechanical, electrical, and chemical energy to remove material from a workpiece. The hardness and shape of the workpiece are no longer a barrier in non-conventional machining. Traditional machining, also known as conventional machining, demands the use of a tool that is harder than the workpiece to be machined. This tool must be introduced to a certain depth into the workpiece. Furthermore, the desired shape is created or generated by the relative motion of the tool and the workpiece. A tool has a fixed number of cutting edges with a specific shape. During a cutting action, observable chips are eliminated, which is called machining allowance. The traditional manufacturing processes (TMP) are difficult to process the hard and brittle material such as welding, casting, metal forming, and powder metallurgy, and traditional machining such as turning, drilling, reaming, boring, milling, broaching, slotting, shaping, and grinding [1–3]. The key drawbacks of TMP are that the cutting tool must be harder than the workpiece material; the tool life is less due to the high wear rate of a tool; poor accuracy and surface finishing; high heat generation due to friction can distort or crack the surface or cause changes to the microstructure of the working material; and the machining of brittle and fragile metal components cannot be achieved without causing distortion [4–6]. A non-traditional machining process is a machining method in which the tool does not make direct contact with the workpiece. Unconventional machining utilizes energy to remove undesirable material from a particular workpiece. However, a new generation of manufacturing processes are not bounded by the above limitations in the machining or manufacturing of hard and brittle materials to produce complicated shapes or profiles. In order to overcome the above drawbacks, newer methods have been developed. These newer manufacturing

DOI: 10.1201/9781003375098-7

processes are called non-traditional or modern or new generation machining/manu-facturing processes. Modern manufacturing processes are classified according to the basic operations of energy. A detailed classification of the processing process according to the type of energy, the metal removal mechanism, resources, energy requirements, etc. [7,8] includes ultrasonic machining process (USMP), abrasive jet machining process (AJMP), abrasive water jet machining process (AWJMP), electron beam machining process (EBMP), laser beam machining process (LBMP), plasma arc machining process (PAMP), electric discharge machining process (EDMP), and electrochemical machining process (ECMP) [9,10]. The manufacturing process is classified into two categories: the conventional manufacturing process and the non-traditional manufacturing or new generation manufacturing process.

7.2   CONVENTIONAL MANUFACTURING PROCESS

7.2.1   WELDING PROCESS

Welding is the process of joining two metals, whether they are similar or dissimilar. It can combine various metals/alloys with or without pressure and with or without filler metal. Heat is used to perform metal fusion. Gas combustion, electric arc, elec-tric resistance, and chemical reactions can all generate heat. As a result, for the vast majority of critical components, it is usually followed by post-weld heat treatment. Welding is a common production and repair process in industry. The technique of welding is used in the construction of ships, pressure vessels, automobile body work, offshore platforms, stadiums, welded pipes, nuclear fuel and explosives sealing, and other structures [5,6].

7.2.2   CASTING PROCESS

Casting is a manufacturing process in which molten metal is poured into a properly constructed chamber. When the metal cools and solidifies, it adopts the shape of a cavity. Casting refers to the cooling of a solid metal. Following that, the required form and size casting are removed from the cavity. Pattern creation, sand prepara-tion, molding, metal melting, pouring into molds, cooling, shakeout, heat treatment, finishing, and inspection are all part of the process [1–3,5]. A casting process can be used for the following applications: aviation, automobiles, trains, shipping, chemical machinery, petroleum, steel and thermal plants, defense (artillery, weapons storage, vehicles, and supporting equipment), hardware (joint plumbing industry, pipe fitting and coupling), and household (kitchen appliances, gardening equipment, furniture, and fittings).

7.2.3   METAL FORMING PROCESS

Metal forming is one of the world's most common manufacturing techniques. During the forming process, the metal is plastically deformed by a force that exceeds the material yield strength, establishing strain hardening. Low yield strength and high ductility are desirable in the material, with the strain rate and lubricants having a significant impact on performance. Certain forming processes are carried out hot because increasing the temperature lowers the yield strength while increasing ductility, making it easier to deform [4–6,11,12]. Forming processes are classified based on the type of force employed as follows: tensile stress forming, combination stress, and bending, shearing, and compressive stress forming.

Processes can also be classified into several forms according to how the force is supplied.

- **Rolling:** The most common type of deformation is rolling. It involves passing metal between two rollers that produce compressive pressures on the metal, lowering its thickness. When producing simple shapes in big quantities, rolling is the most cost-effective method.
- **Forging:** Forging is a metal forming method that employs focused compressive stresses. There are four basic methods in forging: impression die forging, cold forging, open die forging, and seamless rolling ring forging.
- **Extrusion:** Extrusion is a metal forming method in which metal or a work item is forced to flow through a die to shrink its cross section or shape. This procedure is commonly used in the manufacture of pipes and steel rods. Compressive force is applied to extrude the workpiece.
- **Drawing:** Drawing is a metalworking procedure that uses tensile pressures to elongate metal, glass, or plastic. As the material is drawn (pulled), it stretches and becomes thinner, producing the required form and thickness.
- **Sheet metalworking:** Sheet metal forming procedures include applying force to a portion of sheet metal with the objective to change its geometry rather than remove material. The applied force pushes the metal beyond its yield strength, causing it to plastically deform but not fail. It is cold deformed using a punch and die assembly. Because of the use of press machinery, this method is also known as press work.

7.2.4 POWDER METALLURGY

Powder metallurgy is a production method that uses extreme pressure to crush powdered metals and alloys into hardened dies to manufacture extremely accurate products. With the development and application of new advancements, powder metallurgy has emerged as the key method for producing bushings, bearings, gearboxes, and a variety of structural components. The precision and profitability of powder metallurgy rely on the sintering procedure, which heats parts to fuse powder particles. The gaps between the powdered particles are bonded through sintering, which takes place at a temperature that is below the melting point of the main metal [1–4,13]. The basic powder metallurgy steps are as follows:

7.2.4.1 Powder Preparation

The qualities of powdered metallurgical goods are influenced by the features and attributes of the powder. The melt atomization process is one of the methods used to create metal powders. The liquid metal is separated into small droplets that cool and solidify into tiny particles during the process. Although atomization is the most frequently used process for producing powder, other methods such as chemical reduction, electrolytic deposition, grinding, and thermal breakdown are also used. Any metals or alloys can be turned into powder using whatever procedure is employed. The overall compatibility for powder metallurgy technology is investigated and tested prior to combining and blending the powder. Flow rate, density, deformability, and strength are all factors to consider.

7.2.4.2 Mixing and Blending

Powders are mixed and blended with other powders, binders, and lubricants to ensure the finished product has the required properties. Depending on the type of powder metallurgy process utilized and the parameters of the item, blending and mixing can be done dry or wet. The rotating drum, revolving double cone, screw mixer on a drum's interior, and blade mixer on a drum's interior are the four most prevalent blending and mixing methods.

7.2.4.3 Compacting

Compression is the act of squeezing and consolidating a powder mixture into a specific shape or die. When done correctly, this process, known as compaction, enhances the density of the product and minimizes undesirable empty spaces. The resulting form, referred to as a "green compact", signifies the utilization of compaction in component creation. The pressure range employed for compaction typically falls between 80 and 1600 MPa. The optimal amount of compacting pressure varies depending on the specific characteristics of the metal powder involved. For soft powder, the pressure usually ranges from 100 to 350 MPa, while harder and more durable metals such as steel and iron necessitate a pressure range of 400–700 MPa.

7.2.4.4 Sintering

Powder metal sintering is a technique in which particles under pressure chemically attach to each other to form a coherent shape when subjected to high temperatures. The temperature at which the particles are sintered is almost always lower than the melting point of the powder's major component.

7.2.4.5 Finishing Operations

A thoroughly sintered component retains significant porosity (4%–15%). Density is frequently kept low on purpose to maintain interconnected porosity in bearings, filters, acoustic barriers, and battery electrodes. To enhance the properties, however, finishing actions are required:

- Cold restriking, re-sintering, and heat treatment are all options.
- Impregnation with hot oil.
- Metallic infiltration (e.g., Cu for ferrous parts).
- Tighter precision machining

Finishing procedures include the following applications: sizing, coining, infiltration, impregnation, machining, heat treatment, and plating.

7.2.5 CONVENTIONAL MACHINING PROCESS

7.2.5.1 Lathe Machine

A lathe is a tool that turns a work item while it is safely held in a chuck or face plate, or between two rigid and sturdy supports known as centers. The cutting tool is tightly held and maintained in a tool post before being fed against the revolving workpiece. While the workpiece rotates around its own axis, the tool is intended to move parallel to or at an angle to the axis of the material to be cut. As a result, the principal function of a lathe is to remove metal from a workpiece so that it can be shaped and sized. The material from the workpiece is removed in the form of chips [6,7]. A lathe machine can conduct the following operations:

- **Turning:** The removal of metal from the outside diameter of a revolving cylindrical workpiece is known as turning. Turning is a process that is used to smooth out metal and reduce the circumference of a workpiece, usually to a specific size.
- **Taper turning:** Tapered means that the diameter of an item changes consistently from one end to the other. Taper turning is a machining technique that involves gradually decreasing the diameter of one portion of a cylindrical workpiece to another.
- **Facing:** Facing is the removal of metal from a workpiece's end in order to make a flat surface.
- **Drilling:** Drilling is the process of creating a circular hole in a workpiece.
- **Boring:** Boring is the process of expanding a hole when the perfect-size drill is absent. It should be observed, however, that drilling cannot create a hole.
- **Reaming:** Reaming is considered a finishing procedure since only a minor quantity of material is removed throughout the process.
- **Knurling:** Knurling is the procedure of employing a particular knurling tool to emboss a diamond-shaped regular pattern on the surface of a work item.
- **Tapping:** Tapping is a procedure that involves cutting internal threads using a tapping device.
- **Threading:** Threading is simply the creation of a helical groove on a work item. Internal or external cylindrical surfaces can be used to cut threads.
- **Counter-sinking:** For a short distance, the counter-sinking process is used to enlarge the end of a hole and give it a conical form.
- **Counter-boring:** Counter-boring is a procedure that is used to enlarge only a small portion of a hole.

7.2.5.2 Drilling Machine

Using a drill, a spinning cutter, to produce a circular hole in a workpiece is known as drilling. The name given to this device is the drilling machine. Holding the drill in the tailstock and the work in the chuck permits the drill to be used in a lathe. Twist drills are the most common drill. It is the most fundamental and accurate machine in manufacturing. When actuated by moving linearly against the workpiece, a power-driven machine tool holds the drill in its spindle, which revolves at high speeds and makes a hole [4–6]. A lathe machine can perform the following operations:

- **Drilling:** The operation of creating a cylindrical hole in a solid body using rotating equipment is called drilling.
- **Reaming:** The operation of completing a drilled hole using a tool is called reaming.
- **Boring:** Boring enlarges a hole and is used when the proper-size drill is unavailable. It should be emphasized, however, that drilling cannot create a hole.
- **Counter-Boring:** A counter-boring instrument is used to enlarge a drilled hole to a limited length.
- **Counter-sinking:** With a counter-boring tool, counter-sinking enlarges the top end of a drilled hole and molds it into a conical shape.
- **Spot facing:** Spot facing is squaring up the surface at the top of a hole in order to give a proper seat for a bolt head or collar.
- **Tapping:** Tapping is a procedure that involves cutting internal threads using a tapping tool.

7.2.5.3 Shaper Machine

Metal is often removed by a shaper during the forward cutting stroke, but not during the return stroke. A reduced return stroke time will reduce overall machining time. As a result, it should be built in such a way that the ram holding the tool can travel at a significantly slower speed during the forward cutting stroke. The cutting speed varies according to the material and machining conditions; however, during the return stroke, the ram can operate at a quicker rate to decrease idle return time. This is known as the rapid return mechanism. The workpiece is stationary in a shaper machine, while the cutting tool moves back and forth. The cutting tool is secured in the tool post, and the workpiece is securely kept in place on the table. Shapers are usually used with specific attachments to generate flat, smooth surfaces. They can also make internal and exterior keyways, gear racks, spiral grooves, T-slots, and other shapes. The following procedures can be carried out on a shaper: machining of vertical, horizontal, angular, and irregular surfaces; machining splines or cutting gears; and cutting slots, grooves, and keyways [1–4].

7.2.5.4 Planer Machine

A planer machine eliminates unwanted material from an item of metal to provide a flat surface on the workpiece. Unlike the shaper machine, this machine allows for the use of many tools to complete an operation. The work table can be moved, but the machine's tool head remains fixed. The workpiece is secured to the work table, and a single point cutting tool is mounted to the tool head. The machine is turned on, and the work table moves forward. As a result, it slices the material, which is referred to as a cutting stroke. The return stroke is when the work table goes downward without cutting any material [5,7]. Until the power source or other components are updated, the process will continue. Planer machine applications include the following:

- Used on workpieces with flat surfaces.
- Used in cutting angular surfaces.
- Used to make grooves and slots.

7.2.5.5   Milling Machining

Milling is a machining operation in which cutters revolve to remove material from the workpiece in the direction of the tool axis's angle. Milling machines can be used to conduct a wide range of operations and activities on tiny or large items. Milling machining is a typical production procedure used in machine shops and industries to generate high precision products and parts in a wide range of shapes and sizes. Milling machines are also known as multitasking machines (MTMs) because they are multipurpose equipment that can mill and rotate materials. A cutter is mounted on the milling machine, which facilitates the removal of material from the workpiece's surface. When the material has cooled, it is removed from the milling machine. Milling machines are used for machining flat surfaces, slotting, and contoured surfaces. They can also be used to create complex and uneven regions, revolution surfaces, gear cutting, and external and internal threading [1–3].

7.2.5.6   Grinding Machine

Grinding, as with any other machining process, is a metal cutting technique that removes metal in a proportionally smaller volume. An abrasive wheel with several cutting edges is utilized as the cutting tool. A grinding machine performs the grinding action. Grinding is carried out to achieve very high-dimensional precision and a more pleasing look. The grinding process has a precision of 0.000025 mm. Very little material is removed from the piece. The grinding machine operates on the grinding process concept. This is a machining technique that involves finishing work on the workpiece. A solid item made of abrasive particles is moved relative to the workpiece in this operation. As these abrasive particles come into contact with the workpiece due to this relative motion, abrasion occurs and superfluous material is removed from the workpiece. A grinding wheel is typically utilized in this operation since it is made up of abrasive particles. This grinding wheel is driven by an electric motor. The revolutions per minute vary from 150 to 15,000 rpm depending on the machine type, operation, and finishing quality.

- **External cylinder grinding:** External cylinder grinding is the process of grinding tapered and cylindrical surfaces that are situated outside of the cylinder.

- **Internal cylinder grinding:** Internal cylinder grinding is the process of grinding tapered or cylindrical surfaces that are situated inside the cylinder.
- **Horizontal, flat surfaces grinding:** Horizontal, flat surfaces grinding uses a surface grinder. The spindle of the wheel can be vertical or horizontal.
- **Centerless grinding:** In order for the workpiece to move longitudinally, two grinding wheels are positioned parallel to one another at 5°–10° angles. A cylindrical rod is placed between the two grinding wheels, and because of the sloping angle, the workpiece naturally slides between the wheels, resulting in a smooth surface.
- **Form grinding:** This type of grinding process maintains the grinding wheel in a form identical to the completed product. When the task progresses through it, an automated production of the predetermined shape takes place on the workpiece.
- **Wet grinding:** In wet grinding, coolant is continually sprayed to preserve a cold surface, increase the lifespan of the grinding wheel, and provide a good surface finish.
- **Dry grinding:** This type of grinding does not need coolant. Dry grinding is not recommended since it results in an uneven surface finish and increased grinding wheel wear.

7.3   NEW GENERATION MANUFACTURING PROCESSES

7.3.1   ULTRASONIC MACHINING PROCESS

7.3.1.1   Introduction to USMP

The USMP is a non-traditional machining method in which abrasives are used to remove undesirable material from a workpiece. The tool is used to strike or hammer abrasives against the workpiece throughout the machining operation. In this process, a tool oscillating at low amplitude (25–100 μm) and high frequency (15–30 kHz) forces abrasives in the slurry against the work. Ultrasonic waves have a frequency that is greater than the hearing range of 20 kHz. Balamuthia proposed the ultrasonic machining method in 1945. It was created for the purpose of completing the components for electric spark machines. The USMP is composed of a tool produced from a durable and ductile material. The tool oscillates at a high frequency, and abrasive slurry is continually fed between the tool and the workpiece. The impact of the hard abrasive particles fractures the workpiece, removing the tiny particles from the work surface. It consists of an abrasive slurry, a workpiece, a table, a fixture, a cutting tool, a reservoir, a circulating pump, ultrasonic oscillators, leads, an excitation coil, a feed device, a transducer cone, an ultrasonic transducer, a connecting member, and a tool holder [8–15]. To convert electrical energy at low and high frequencies, an ultrasonic generator and a power amplifier, also known as a power generator, are employed. The transducer is built of magneto strictive material and is composed of a stack of nickel laminations wound with a coil. The primary job of the transducer is to convert electrical energy into mechanical energy. Stainless steels and low carbon steels are commonly used as tool materials. The tool is often soldered or physically connected to the transducer through a metal holder. Tool holders are typically conical or cylindrical in form. Tool holders are made of titanium alloys, stainless steel, Monel, and aluminum. Abrasive slurry, which is typically a combination of abrasive grains and water in a predetermined proportion (21%–30%), is forced under pressure through a gap of 0.02–0.1 mm between the tool and the workpiece. Aluminum oxide, boron carbide, silicon carbide, and diamond are some of the most regularly used abrasives. Boron carbide (B4C) is utilized as a raw slurry due to its abrasive and quick-cutting qualities [15–18].

7.3.1.2 Working Principle of USMP

Ultrasonic machining is a mechanical machining method that removes material from a workpiece by microchipping and surface erosion with abrasive grain in the slurry. A tool oscillating normal to the work surface at high frequency moves the abrasive grains between the tool and the workpiece at ultrasonic speed (20 kHz). As the abrasive-loaded slurry abrades the work materials, the tool generates a reverse image in the workpiece [19–21]. Figure 7.1 is a schematic representation of the ultrasonic machining system. The transducer produces high-frequency vibrations with an amplitude of approximately 0.02 mm when high-frequency power is applied. Using a coupler, referred to as a tool holder, the vibration is transferred to the appropriately shaped tool. The soft construction of the tool allows for the incorporation of abrasive particles, enabling it to function as a multipoint tool with embedded abrasives. The tool shape is a near match to the final surface to be created. As the metal removal from the workpiece advances, the oscillating tool is forced against the workpiece and supplied continually. A servomechanism controls the feed so that a consistent distance between the tool and the workpiece is maintained. While the rate of metal removal is somewhat slow, it is the only technique to build economically intricate cavities in brittle and delicate materials without the risk of facture [18–22]. Table 7.1 summarizes the main points of the USMP.

7.3.2 Abrasive Jet Machining Process

7.3.2.1 Introduction to AJMP

The abrasive jet machine uses a high-velocity carrier gas and abrasive combination to remove undesirable material. It is an unconventional machine tool in which there is no physical connection between the tool and the workpiece. Surface cleaning, trimming, engraving, coating removal, deflashing, drilling, parting, deburring, glass frosting, ceramic abrading, and additional applications require this technique [18,21,22]. It has numerous important advantages over other non-traditional cutting processes, most notably good machining flexibility and low substrate stresses. Franz exhibited this new process for cutting laminated paper tubes for the first time in 1968. In 1983, a commercial system for abrasive jet machining was created. Dr John Olsen, a water jet pioneer, began investigating the notion of abrasive jet cutting as a realistic option for typical machine shops in the early 1990s. The overall goal was to develop a product that would lessen the skill, noise, and dust associated with abrasive jets at the time. In AJMP, a mechanical energy–based advanced machining technique,

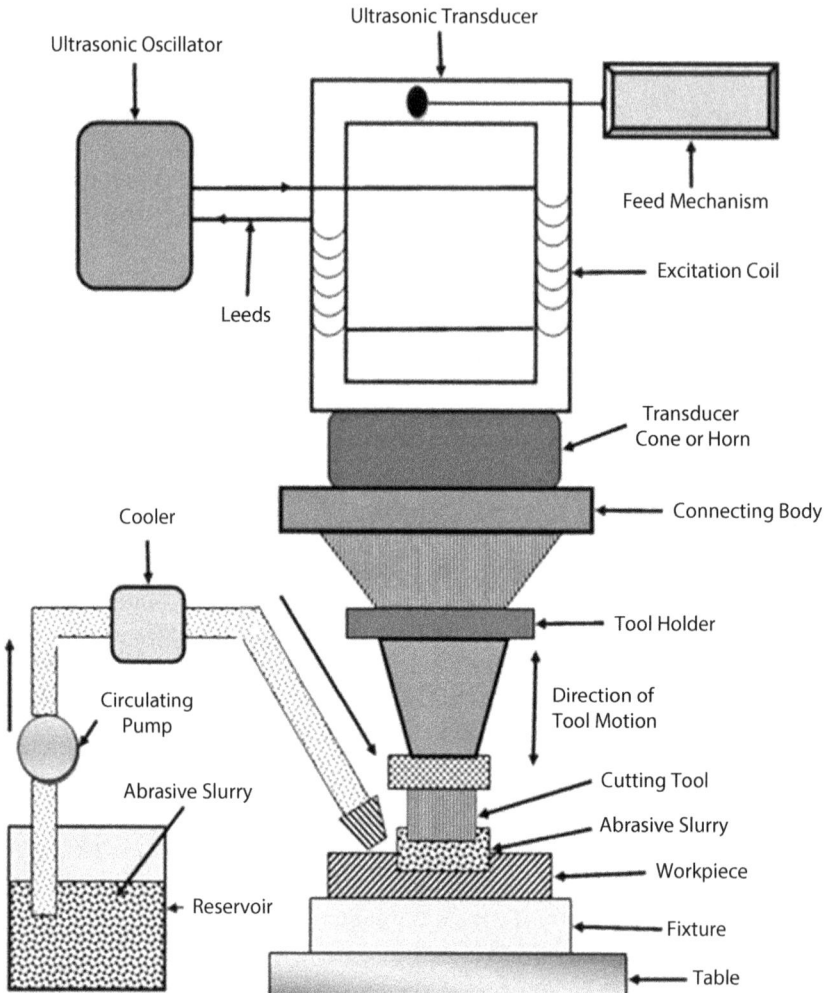

FIGURE 7.1 Schematic diagram of the ultrasonic machining process.

material is eliminated from a work surface by impact erosion using a high-velocity jet of abrasives. The abrasive jet is created by rapidly accelerating tiny abrasive particles/powder in a high-pressure gas (carrier gas). A nozzle is used to transform the pressure energy into kinetic energy and to guide the jet at a certain angle toward the work surface (impingement angle). When abrasive particles/powder collide, they gradually remove material by erosion and, to a lesser extent, brittle fracture. The hardened nozzles of this process are often composed of tungsten carbide or sapphire. Glass beads, silicon carbide, dolomite, sodium bicarbonate, and aluminum oxide are common abrasive powders. The AJMP consists of (i) a compressor to compress air or gas (carbon dioxide and nitrogen) to high pressure; (ii) an abrasive powder; (iii) a

TABLE 7.1
Key Points of USMP

1.	Material removal technique	Slurry of small abrasive particles is forced on the workpiece by a vibrating tool
2.	Medium	Slurry
3.	Abrasives	B4C, SiC, Al$_2$O$_3$, diamond dust
4.	Amplitude	Range 25–100 μm
5.	Vibration frequency	Range 15–30 kHz
6.	Tool material	Stainless steel and low carbon steel
7.	Work material	Tungsten carbide, ceramics, glass and quartz, ferrite, diamonds, and alumina
8.	Gap	25–40 μm
9.	Material removal	14 mm^3/s
10.	Surface finish	0.2–0.7 μm
11.	Capital cost	High
12.	Critical parameters	Amplitude, frequency, grit size, feed force, slurry concentration, tool material, slurry viscosity, abrasive material
13.	Applications	Precise machining, engraving, profiling the hole, trepanning, coining, slicing, threading, grinding of brittle material, and broaching of hard metal
14.	Advantages	• Burr-less process • It can be employed as a machinable brittle, hard, and non-conductive material • No thermal effect on the machined items • Low cost of metal removal • Capability of drilling circular and non-circular
15.	Disadvantages/ limitations	• Low MRR • High tooling cost • Not suitable for heavy stock removal • Tool wear rate is high due to abrasive particle • Depth of holes and cavities small • Inability to machine soft material • Periodic replacement of abrasive slurry

mixer to mix air and abrasive powder; and (iv) a nozzle to convert pressure energy into kinetic energy and to focus the air with fine abrasive particles on the work surface. The gas/air pressure affects the rate of material removal. It grows when the gas pressure rises. The type, size, and form of abrasive particles affect the cutting rate. As their cutting effectiveness deteriorates after one usage, abrasive granules are never reused [21–26].

7.3.2.2 Working Principle of AJMP

The AJMP works by using a nozzle on the workpiece to direct a high-velocity flow of compressed abrasive particles delivered by a high-pressure gas. When fine abrasive particles impact a workpiece at very high speed, they erode the metal, removing it. The pressure energy of the jet stream is transformed into kinetic energy, which produces the high-velocity jet. Through micro-cutting and brittle fracture, the high-velocity tiny abrasive particles remove the material from the work surface. The head of the nozzle can be straight or at an angle, and its cross section can be round or rectangular. It is intended to lose the least amount of pressure feasible as a result of friction and bends. When the divergence of the jet stream rises, more irregular cutting and increased incorrectness occur [23–26]. Figure 7.2 depicts a schematic representation of the AJMP.

7.3.2.3 Parts/Components of AJMP

The AJMP is a contemporary (mechanical energy–based) machining method that erodes material from a work surface using a high-velocity jet of fine grain abrasives combined with a compressed carrier gas. An air compressor with a drier and cleaner, an abrasive feeder and vibratory mixing chamber, a nozzle, a well-ventilated machining chamber, and several gauges for measuring pressure, flow rate, and so on are all part of the AJMP system. A thorough explanation of such a configuration is provided below.

1. **Gas propulsion system:** Delivers clean, dry air, carbon dioxide (CO_2), and nitrogen (N_2) to propel the abrasive fine particles. Gas can be delivered by a cylinder or a compressor. An air filter cum drier should be used in the case of a compressor to keep abrasive particles/powder from being contaminated with water (H_2O) or oil. Gas should be non-toxic, cheap, and abundantly

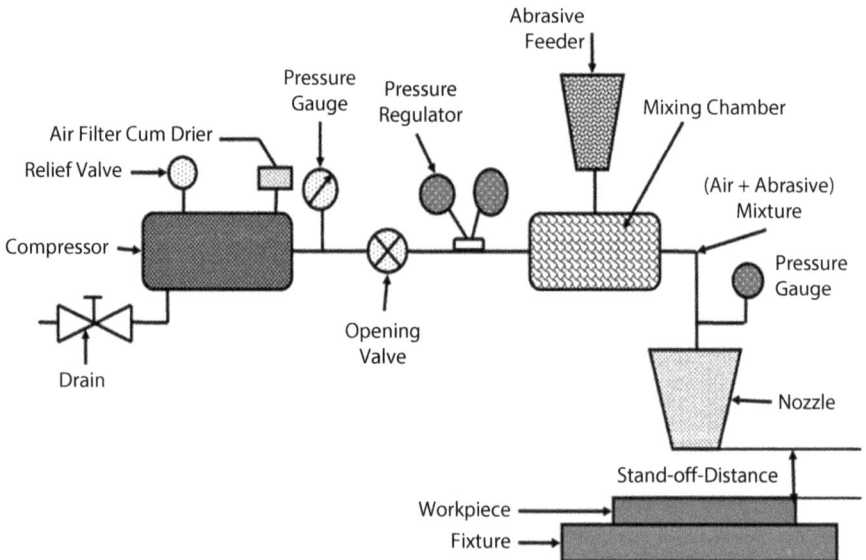

FIGURE 7.2   Schematic diagram of the AJMP setup.

available. When ejected from the nozzle into the environment, it should not be too dispersed. The propellant consumption is of the order of 0.008 m³/min at a nozzle pressure of 5.0 bar, and the abrasive flow rate ranges from 2.0 to 4.0 g/min for fine machining and 10–20 g/min for cutting activities.

2. **Abrasive feeder:** The necessary quantity of abrasive powder is delivered by the abrasive feeder. Abrasive particles are added via a filter into the mixing chamber before the filleted propellant is inserted. The mixing ratio is controlled by the amplitude of the strainer's oscillation, which is adjusted to vibrate at 45–65 Hz. The carrier gas (air, CO_2, or N_2) transports the abrasive particles to a mixing chamber. The nozzle moves closer to the abrasive air mixture. Since the mixture is directed at the surface of the workpiece by the AJM nozzle, it travels at a high velocity.

3. **Machining chamber:** It is securely enclosed to prevent dangerously high levels of abrasive material concentration surrounding the operating chamber. The machining chamber features a dust collection system with a vacuum. The dust collection system requires extra care when working with hazardous compounds (e.g., beryllium).

4. **Air compressor:** Normally, air is taken directly from the atmosphere, dry and dust-free, and is then compressed to a high pressure. Depending on the compressor's capacity and the needed jet velocity, the carrier gas pressure is maintained at between 15 and 20 bar. In unusual circumstances, the use of commercially pure nitrogen or carbon dioxide gas can also improve outcomes.

5. **Filter regulator lubricating (FRL) unit:** High-pressure air is sent through a filter regulator lubricating unit to remove any suspended particles such as dust or oil. An FRL device is frequently mounted on a compressor to keep the carrier gas dry and dust-free. The presence of steam in compressed gas is especially undesirable because it can congeal and generate abrasive agglomeration during pipeline transmission.

6. **Pressure gauge:** A pressure gauge is a tool for determining the level of pressure in a fluid (air or gas). This tool assesses if the necessary quantity of pressure is present. The function of this device is usually to check the pressure.

7. **Air filter and drier:** Its major purpose is to purify the air or gas. Compressed air dryers are sophisticated filtration appliances created to remove water from compressed air.

8. **Relief valve:** Relief valves are commonly used to regulate the pressure in a system, which is most common in fluid or compressed air systems. The opening of these valves is proportional to the rise in a system's pressure.

9. **Servo controller:** A servomechanism is commonly used to control the movement of a work table. Control is easy, accurate, and precise, making it ideal for cutting difficult profiles and shapes.

10. **Abrasive drainage system:** The abrasive particles delivered by the high-pressure air stream must be appropriately disposed of or they can build and cause additional issues. A two-way sloping tunnel provides for particle

drainage. The route is built of an aluminum sheet and has a slope in both the x and y axes. It is curved such that the particles proceed to a corner and exit.

11. **Work holding device:** The most frequent work components in the AJM machining process are glass sheets, glass fiber sheets, and ceramic slabs because of their fragility. Due to their tendency to embed in the work material, ductile materials have a substantially lower rate of material removal. To hold the workpiece in place, L-shaped angle plates with holes drilled along their length are employed. In order to secure the plate and the workpiece together to form a permanent unit, the workpiece also has holes drilled into it. The L-plates are secured to the mild steel box. The major benefit of this kind of work holding system is its ability to handle various workpiece forms by simply adding new plates.

12. **Opening and closing valve:** For ease of opening and closing, the chamber hinge joint is employed. This method can be used to remove the glass fiber front panel. The hinges are typically composed of mild steel and are low weight. Nuts and bolts hold it to the mild steel box. To fasten the wall, magnetic clamps are employed. Typically, they are constructed of a magnet and a steel plate, with the magnet attached to a fixed wall and the metal plate attached to a moveable door or plate. When they come into contact, the magnet tightly grabs them. A rubber strip is added to the plate's edge to prevent (air + abrasives) leakage via the area between the metal and glass fiber.

13. **Nozzle:** As the gas–abrasive combination approaches a nozzle, hydraulic energy (pressure) is transformed into kinetic energy, producing a high-velocity abrasive jet. A certain distance between the nozzle and the work surface is maintained. Consequently, the gap is also known as the stand-off distance (SOD), and it is an important factor affecting the quality of a machining process. The inclination angle, or the angle between the work surface and the jet axis, is also controlled by nozzle alignment. Moreover, the nozzle is mounted on a slider to deliver the required movements based on the required cut profile. A typical AJM nozzle is composed of sapphire or tungsten carbide (WC), both of which have a long lifespan of 20–30 h and 300 h, respectively. The AJM nozzle can have a straight or right-angled head, and it can have a rectangular or circular cross section. Its construction minimizes pressure (P) loss caused by bends, friction, and other elements. Wear of a nozzle causes the jet stream to diverge more, which causes more stray cutting and a worse performance. Here, the fundamental functions of nozzles in abrasive jet machining are described.

    • Delivering a high-velocity jet: The pressure of the gas–abrasive combination is high (15–20 bar) but the velocity is low at the nozzle's input. The static enthalpy (pressure) of this mixture is converted into kinetic energy when it passes through the nozzle, producing a high-velocity jet.

    • Controlling jet diameter: The internal shape of the nozzles and the exit diameter both influence the diameter of the abrasive jet. A bigger diameter decreases jet velocity but increases the jet cross-sectional area, and vice versa, at a constant carrier compressed gas and mixing ratio.

- Regulating impingement angle: The AJM nozzle adjusts the impingement angle in addition to producing a high-velocity jet. It is defined as the angle created between the work surface and the jet axis. Its quantity can potentially range from 0° to 90°; however, in reality, it is maintained at between 60° and 90°. It affects the material removal rate and machining accuracy.

14. **Work material:** It is ideally suited to working with hard, brittle, and glass sheets.

15. **Modern accessories and controlling:** The modern abrasive jet machining system includes a pneumatic controller, a pneumatic timer, many sensors, and computer numerical control system. The length of the blast can be accurately controlled using one of these timers. At the entry and exit of the nozzle, sensors can be employed to measure the pressure and flow rate. These sensor readings aid in preventative maintenance, the prompt replacement of worn-out nozzles, and the prevention of machining errors. Moreover, a CNC system with a stepper motor is employed to precisely position the nozzle and manage numerous processes. Table 7.2 summarizes the main points of the AJMP.

7.3.3 ABRASIVE WATER JET MACHINING

7.3.3.1 Introduction to AWJMP

In a complex water jet machining procedure, abrasives are used as the medium. To overcome the drawback of water jet machining, abrasive particles are combined with water and then ejected from a nozzle at very high velocity to remove material from a workpiece's surface in order to precisely shape and enhance it to the desired dimensions [1,16,17,23,24].

7.3.3.2 Working of AWJMP

From the reservoir to the drain system, the entire system must receive water from the reservoir. The process of moving water begins with the hydraulic intensifier, where its pressure is raised before being sent to the accumulator for short-term storage. Water pressure and flow direction are controlled in the system via control valves. Then, in the mixing chamber, water is mixed 30:70 with abrasive particles. The hard work materials can fracture due to plastic deformation when these high-velocity abrasive particles impact them. The amount of material removed increases as the abrasive ratio increases, but the flow characteristics of the liquid mixture deteriorate. As a result, the ideal abrasive percentage is between 40% and 60%. It was primarily designed for cutting complicated forms out of granite and marble [11,23–26]. Figure 7.3 depicts the operations of the AWJMP.

TABLE 7.2
Key Points of AJMP

1.	Working principle	It works by combining a high-pressure stream of fine-grained abrasives (10–40 μm) with air or another carrier gas. This stream is directed onto the work area using a suitable nozzle. The rate of carrying gas or air can range from 200 to 400 m/s.
2.	How it works	Material removed by erosive/chipping action
3.	Medium	Air, CO_2, N_2, and He
4.	Abrasives	• SiC, Al_2O_3, $NaHCO_3$, glass beads, and dolomite • Shape: Irregular and regular • Size range: 10–50 μm • Mass flow range: 2–20 g/min • Reuse: Not recommended (contamination with chips, block nozzle passage), lower cutting ability
5.	Machining chamber	• Well closed: Abrasive particles contents should be below harmful, vacuum dust collector
6.	Abrasive feeder	• Delivers the necessary amount of abrasive flow at the optimal pressure and velocity
7.	Velocity of jet	Range 150–300 m/min
8.	Nozzle pressure	Range 2–8 kg/cm²
9.	Nozzle cross section	Rectangular, circular
10.	Mixing ratio	Volume flow rate of abrasive/volume flow rate of carrier gas
11.	Nozzle size	Range 0.07–0.40 mm
12.	Material of nozzle	Tungsten carbide (WC) and sapphire
13.	Nozzle life	Range 12–300 h
14.	Stand-off distance	Range 0.25–15 mm (8 mm generally) • Low SOD: Higher accuracy, low kerf width, and low taper
15.	Work material	• Non-metals, e.g., granites, glass, and ceramics • Metals and alloys of hard materials, e.g., germanium and silicon
16.	Capabilities	• Low MRR: 15 mm³ • Intricate shape • Narrow slots (0.120–0.250 mm) • Low tolerance (±012–0.25 mm) • Minimization of taper: Angle of nozzle wrt w/p • Almost no surface damage
17.	Applications	• Drilling • Cutting • Polishing • Deburring • Etching • Cleaning
18.	Advantages	• No thermal damage • Suitable for non-conductive material • Removal of deposit (oxide and coating)
19.	Limitations	• Low material removal rate • Tapered and stray cutting • Abrasive cannot be reused • Not appropriate for materials that are ductile and soft • Limited nozzle life

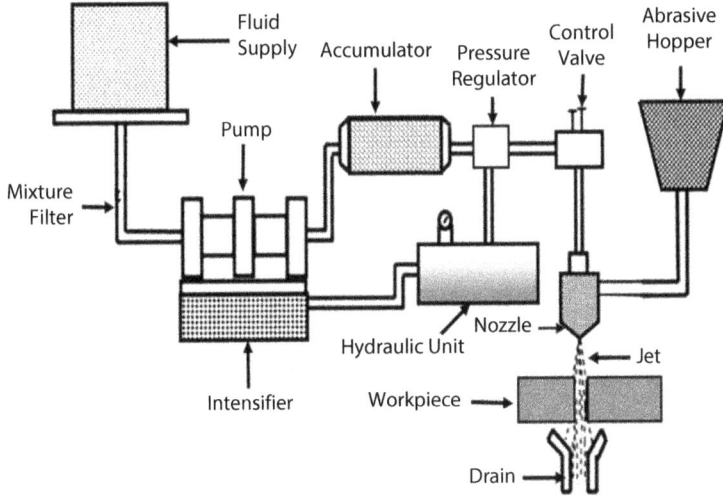

FIGURE 7.3 Line diagram of abrasive water jet machining.

7.3.3.3 Parts of AWJM

The main elements of the abrasive water jet machining process are listed below. Many elements of the AWJM are also included in the setup since the AWJM process is a progression of the WJM process.

1. **Reservoir:** The reservoir serves as a basic component for storing water that is distributed to all other components of the system to ensure appropriate operation.
2. **Hydraulic pump:** The hydraulic pump is commonly employed to transport fluid from one point to another; however, in this case, it acts as a link between the reservoir and the intensifier. Water is drawn from the reservoir and transported to the intensifier using this device.
3. **Hydraulic intensifier:** The hydraulic intensifier is linked to the hydraulic pump and is used to increase the pressure of the water.
4. **Accumulator:** When there is a pressure drop or high-pressure water is required, the accumulator is employed to temporarily store water and deliver it to the system.
5. **Control valves:** The duty of the pressure control valves is to govern both the hydraulic fluid and flow direction. A flow regulator valve regulates the flow of water into the system.
6. **Flow regulator:** The flow regulator, as the name implies, controls the flow of water from the pressure regulator to the nozzle.
7. **Abrasive tank:** This procedure requires the use of abrasive particles. Water jet machining is used to cut soft materials; however, water alone is insufficient to cut hard materials. Abrasives are used in addition to water to allow

the substance to cut quicker. Aluminum oxides, sand, garnet, glass particles, and other abrasive particles are often employed.

8. **Mixing chamber:** Only water is utilized in the WJMP, whereas 30% abrasive particles and 70% water are combined in the mixing chamber during the abrasive water jet machining process.

9. **Nozzle:** The nozzle's job is to convert high-pressure water into kinetic energy (KE), and this KE grows as the nozzle area shrinks. This high KE water, together with abrasive particles, impinges on the workpiece's surface to obtain the appropriate shape and size at a quicker pace.

10. **Drain system:** The major function of the drain is to collect water that runs out of the work area and send it to storage via the pump and filter.

7.3.3.4 Parameters of AWJM

The conditions of the abrasive water jet machining processes are as follows:

1. **Abrasive particle size:** Abrasive particles between 100 and 150 grit are most commonly used in AWJM.

2. **Abrasive materials:** The three most often used abrasive materials are silica, silicon carbide, and garnet (which is 30% more effective than sand).

3. **Traverse rate:** As the traverse rate increases, the depth of cut decreases. The traverse speed is kept between 100 mm/min and 5 m/min.

4. **Stand-off distance:** When the stand-off distance increases, the cut's depth reduces. The stand-off distance is controlled in the 1–2 mm.

5. **Depth of cut:** The cut thickness is restricted to 1–250 mm.

Table 7.3 summarizes the main points of the AWJMP.

7.3.4 ELECTRON BEAM MACHINING PROCESS

7.3.4.1 Introduction to EBMP

A beam of extremely fast electrons is directed at a workpiece as part of this machining technique to remove material from it. Electron beam machining uses high-velocity electrons as its energy source. This approach uses a high voltage direct current (DC) power source to generate highly energetic electrons. It is an unconventional technique for removing material from a workpiece. For process efficiency and to reduce undesired electron scattering, electron beam machining is often carried out in a vacuum chamber. The micro-cutting of metal is the perfect application for this technique [20–22]. A variety of metals can also be drilled using the EBMP.

7.3.4.2 Principle of EBMP

High-velocity electron beams collide with objects converting the kinetic energy of the electron beams into heat. The workpiece material is taken from it as a result of the concentrated heat raising the temperature and vaporizing a small amount of the material. The entire process takes place inside a vacuum chamber because outside of it, the electron would collide with air molecules and lose energy. A tungsten filament

TABLE 7.3

Key Points of AWJMP

1.	Working principle	AWJM employs a high-pressure jet of water (usually 200–300 MPa) in suspended with an abrasive powder. The jet is aimed at the workpiece, which moves as the cutting process advances. The procedure is primarily intended for cutting holes or forms in sheet material.
2.	How it works	Material removed by erosive/chipping action
3.	Medium	Water
4.	Abrasives	• Garnet, silica sand, aluminum oxide, olivine, and silicon carbide
5.	Machining chamber	• Well closed: Abrasive particles contents should be below harmful, vacuum dust collector
6.	Abrasive feeder	• Delivers the necessary amount of abrasive flow at the optimal pressure and velocity
7.	Parameters	• Abrasive flow rate, abrasive particle size, nozzle diameter, water flow rate, water pressure, traverse rate, stand-off distance, abrasive materials, depth of cut, material properties
8.	Material of nozzle	Tungsten carbide (WC)
9.	Stand-off distance	• 1–2 mm
10.	Depth of cut	• Range 1–250 mm
11.	Work material	• Aluminum, hardened tool steel, titanium, copper, brass, or any other exotic metals
12.	Applications	• The AWJM method is used in the aerospace sectors to create parts, e.g., engine components (made of heat-resistant alloys, titanium, and aluminum), aluminum body parts, and titanium bodies for military aircraft • Cut any metal, including aluminum, titanium, hardened tool steel, copper, brass, or any other rare metals
13.	Advantages	• No heat-affected zone • Eliminates thermal distortion • No cutter-induced distortion • Little tooling demands • Workpieces are cut with minimal force • No cutting dross or slag • Limited tooling requirements • Average finish: 125–250 μm
14.	Limitations	• Low material removal rate • Tapered and stray cutting • High startup costs and noise levels • Abrasive cannot be reused • Limited nozzle life • Lower surface finish

wire warms up to 2500°C when a high voltage DC source is attached to an electron gun. The tungsten filament emits electrons as a result of the high temperature. These electrons are directed downward by a grid cup and are driven toward the anode. The anode is exposed to voltages ranging from 50 to 200 kV, which accelerates the

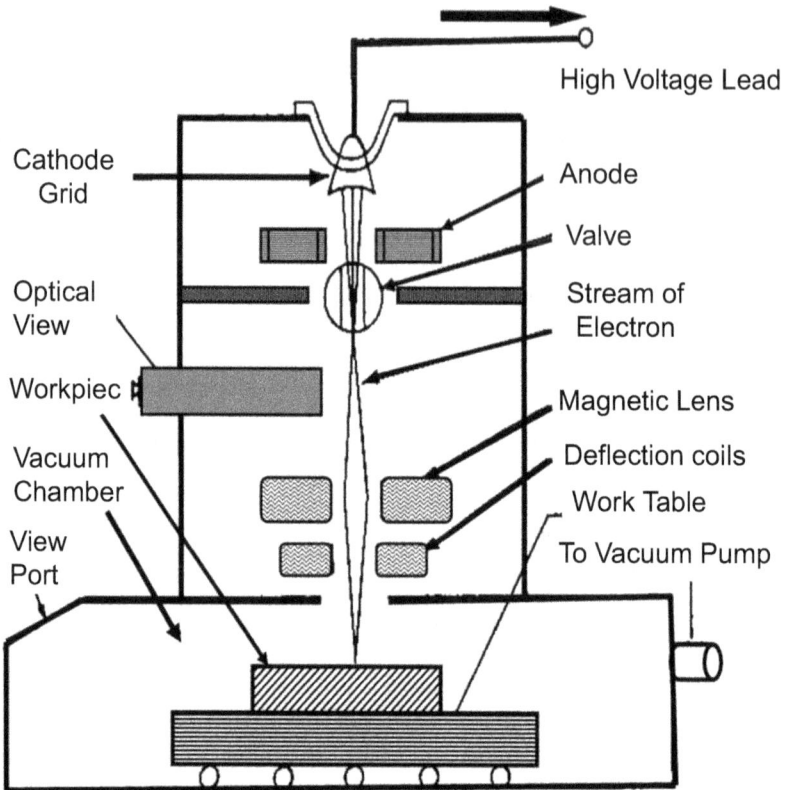

FIGURE 7.4 Schematic diagram of the electron beam machining system.

electrons travelling through it to speeds of as much as half the speed of light (1.6 × 10^8 m/s). These electrons continue moving at high speed until they collide with the workpiece. The fact that electrons can cross the vacuum makes it conceivable. After exiting the anode, this high-velocity electron beam travels via the tungsten diaphragm and the electromagnetic focusing lens. Using the focusing lens, the electron beam is directed to the proper area of the workpiece. The kinetic energy of the fast-moving electrons in an electron beam is swiftly converted into heat energy when it strikes a workpiece. The high-intensity heat vaporizes and melts the work material at the point of beam contact. It takes a few microseconds for the material to melt and evaporate on contact since the power density is very high (about 6500 billion W/mm^2). This procedure is carried out in brief, repetitive pulses. The pulse frequency can range between 1 and 16,000 Hz, while the length can range between 4 and 65,000 μs. The cutting operation can be carried indefinitely by alternately concentrating and shutting off the electron beam [9,10,14,27]. The equipment always has a proper monitoring device. As a result, the operator can easily monitor the progress of the machining process. Figure 7.4 depicts the EBMP working principle.

7.3.4.3   Parts of EBMP

1. **Electron gun:** This is the most crucial factor in terms of electron beam machining. It generates an electron beam that is used to scrape away material from the workpiece. The electron gun's cathode comprises filaments of tantalum or tungsten.
2. **Bias grid:** The electron flow produced by an electron gun is controlled using the bias grid.
3. **Anode:** The anode is employed to accelerate electrons to very high speeds.
4. **Tungsten filament:** This is connected to the negative terminal of the DC power supply and acts as the cathode.
5. **Focusing lens:** The electron beam is shrunk to a 0.01–0.02 mm diameter cross-sectional area by the focusing lens, which concentrates electrons on a small area.
6. **Magnetic lens:** A magnet is used to create a magnetic lens. The main purpose of this magnetic lens is to concentrate an electron beam, similar to an optical lens.
7. **Aperture:** The aperture is similar to a camera's aperture; however, its function is different. The aperture is used to capture errant electrons, permitting only a focused and concentrated beam of electron density to pass through.
8. **Electromagnetic lens:** The electromagnetic lens is used to focus the electron beam on the workpiece.
9. **Diffusion pump:** A diffusion pump is required to keep the electron beam chamber vacuum.
10. **Deflector coil:** The electron beam is directed to various locations on the workpiece by the electromagnetic deflector coil. Also, the cut route can be managed using it.
11. **Optical viewing system:** The operator uses an optical viewing system to determine if the process is or is not in control. A telescope and lighting setup make up this optical viewing setup.
12. **Slotted disc:** In order to prevent vapor and fumes from obstructing the optical windows of the electron beam gun, the vapor and fumes produced when cutting the workpiece with an electron beam are removed using a slotted disc. The electron beam and the slotted disc are synchronized.

Table 7.4 summarizes the keys points of the EBMP.

7.3.5   LASER BEAM MACHINING PROCESS

7.3.5.1   Introduction to LBMP

The LBMP is a thermal machining technology that generates heat and removes material from a workpiece using a laser beam. Metal is removed from the workpiece by melting and vaporizing metal particles in a controlled manner from its surface using laser heat. Laser beam machining is frequently utilized in sheet metal fabrication and hole drilling. It is a non-traditional machining method that employs tools.

TABLE 7.4
Key Points of EBMP

1.	Working principle	The workpieces are impacted by high-velocity electron beams, which heat up as a result of their kinetic energy. The removal of materials from the workpiece is caused by the concentric heat, which raises the temperature of the workpiece's material and somewhat evaporates it.
2.	Accelerating voltage	Range 50–200 kV
3.	Beam current	Range 100–1000 μA
4.	Electron velocity	1.6×10^8 m/s
5.	Medium	Vacuum (10^{-5} to 10^{-6} mm of Hg)
6.	Power density	6500 billion W/mm^2
7.	Workpiece material	All materials
8.	Depth of cut	Up to 6.50 mm
9.	MRR	Up to 40.0 mm^3/s
10.	Specific power consumption	Range 0.5–50 kW
11.	Applications	Precise hole, cutting, micro-drilling, milling, synthetic jewels drilling
12.	Advantages	• Produce very small size holes of any shape • Accurate and precise hole • Machining any type of material • No cutting force b/w tool and workpiece • No tool wear and quicker process, i.e., one hole in 1 s
13.	Disadvantages	• Higher power consumption and investment cost • Low MRR • Skilled operator required • Workpiece size limited due to vacuum chamber

Laser machining is generally used for cutting and drilling. This machining procedure can be used to manufacture both metallic and non-metallic workpieces [27–29].

7.3.5.2 Principle of LBMP

LASER stands for light amplification by stimulated emission of radiation. When electrons in an atom are exposed to an external energy source, they absorb that energy. These electrons move from their initial energy level to a higher energy level by absorbing the energy. The electron releases the absorbed energy in the form of light photons; however, atoms are not in a stable condition, thus it returns to its initial state. Spontaneous emission is the term used to describe this photon emission by electrons. A higher energy level atom will release twice as much energy if it absorbs energy while at that level. Atomic energy has the same frequency and wavelengths as the exciting source. This is the fundamental principle governing laser operation. When a laser material is exposed to an energy source, it absorbs some of the energy and releases it when it reaches its absorption limit. As a result, the extremely magnified light generated is known as laser. The fundamental laser idea serves as the foundation for the laser machining process. This type of machining

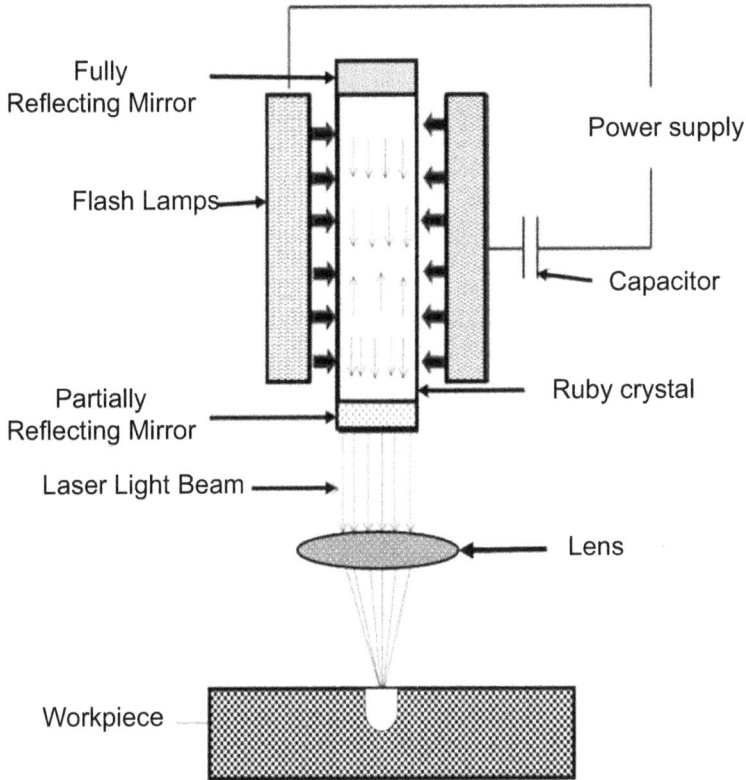

FIGURE 7.5   Schematic diagram of the LBMP.

uses a monochromatic, high-intensity laser beam, which can cut through any kind of material, even non-metals. Diamond, the hardest material that is currently accessible, can be cut and removed with laser machining [10,14,19,20,30,31]. Figure 7.5 depicts the operations of the LBMP.

7.3.5.3   Main Parts of LBMP

The different major components utilized in laser beam machining are as follows:

1. **A pumping medium:** A medium with a significant number of atoms is required. Lasers are made from the atoms of the medium.
2. **Flash tube/flash lamp:** Atoms require energy to excite their electrons, which is provided by the flash tube or flash lamp.
3. **Power supply:** A high voltage is needed for lasers. In order for the electron to leave, the system needs electricity. The electron becomes excited when power is added, indicating that it is prepared to perform its role.
4. **Capacitor:** The laser beam machine is operated in pulse mode by a capacitor. It is used to store and release charge. It is used during the flashing procedure in this case.

TABLE 7.5

Key Points of LBMP

1.	Working principle	A laser beam is used to cut both metallic and non-metallic materials using the LBM technique. In this method, the workpiece is illuminated by a powerful laser beam, and the heat energy of the laser is transferred to the surface of the object.
2.	Mechanism	Material removed by melting and evaporating
3.	Process parameters	Laser power, cutting speed, feed rate and depth of cut, spot diameter of the laser beam
4.	Characteristics	Directionality, monochromaticity, brightness, coherence
5.	Laser medium	He-Ne, Ar, and carbon dioxide
6.	Workpiece material	Metallic and non-metallic
7.	Flash lamps (tube)	More energy is released as one of the primary purposes of flash tubes (used xenon tube)
8.	Material removal rate	5–50 mm^3/min
9.	Types of laser used	Solid-state lasers such as Nd:YAG and YAG lasers, femtosecond lasers, and gas lasers such as CO_2 and excimer lasers
10.	Applications	• Used in the automotive, shipbuilding, aircraft, steel, electronics, and medical sectors for precision machining of complicated parts. It also includes welding, cladding, marking, surface treatment, drilling, and cutting.
11.	Advantages	• Produce very small-size holes of any shape • Accurate and precise hole • Used for soft material, e.g., rubber and plastics • No cutting force between tool and workpiece • No tool wear
12.	Disadvantages	• Higher initial cost and low MRR • Short life of flash lamp • Inaccurately shaped and angled machined holes • Skilled operator required • Unable to drill to deep holes • Thermal damage, heat affected zone, recast layer • Development of a tapering impression as well as dross and splatter

5. **Reflecting mirror:** Two main forms of mirrors are employed: one is fully reflective and the other is somewhat reflective. One end is retained as a fully reflecting mirror, and the other end as a partially reflecting mirror. The side with the partly reflecting mirror is where the laser beams emerge.

6. **Ruby crystal:** The function of a ruby crystal is to convert higher energy into a light beam.

7. **Focusing lens:** The primary function of a focusing lens is to focus the entire light beam and impinge on the workpiece.

Table 7.5 summarizes the keys points of the LBMP.

7.3.6 PLASMA ARC MACHINING

7.3.6.1 Introduction to PAM

Plasma arc machining is an approach to material removal from a workpiece. To melt and remove material from a workpiece using this technique, a high-velocity jet of hot gas is used. This heated gas flows quickly and is also known as a plasma jet. When a gas or air is heated to over 5000°C, it begins to ionize into positive ions, negative ions, and neutral ions. Ionization boosts the temperature of a gas or air to between 11,000 and 28,000°C, and the ionized gas is known as plasma. The arc heats the gas or air, and the plasma that forms as a result of the heated gas is used to remove material from the workpiece. The entire procedure is known as plasma arc machining. With this approach, high-velocity, high-temperature air is used to melt material off the workpiece. The metal chosen as the workpiece affects the gas employed in plasma arc machining. PAM can cut steel alloys, aluminum, stainless steel, copper, nickel, and cast iron [17,21,22].

7.3.6.2 Principle of Plasma Arc Machining

In a plasma torch, known as the gun or plasmatron, a volume of gas, such as H_2, N_2, or O_2, is driven through a small chamber in which a high-frequency spark (arc) is maintained between the tungsten electrode (cathode) and the copper nozzle (anode), both of which are water cooled. Some torches incorporate an inert gas flow encircling the primary flame to shield the gas from the environment. Arc-generated, high-velocity electrons collide with gas molecules, causing diatomic molecules of the gas to dissociate, resulting in atom dissociation and the release of massive amounts of thermal energy. To stabilize the arc, plasma-forming gas is pumped into the nozzle duct of the torch. The gas is heated in the compressed zone of the nozzle duct, resulting in a virtually high exit gas velocity and a high core temperature of up to 16,000°C. The relative plasma jet melts the workpiece material, and the molten metal is effectively blasted away by the high-velocity gas stream. The work material, thickness, and cutting speed all influence the depth of the heat-affected zone. At high cutting speeds, the heat-affected zone on a workpiece with a thickness of 25 mm is roughly 4 mm [9,10,14,21,22,27]. Figure 7.6 depicts the working principle of the plasma arc machining process.

7.3.6.3 Components of PAMP

1. **Plasma gun:** Various gases, such as argon, nitrogen, hydrogen, or a combination of these gases are used to create plasma. Inside the plasma gun's chamber is a tungsten electrode. The electrode is exposed to negative polarity, and the gun's nozzle is exposed to positive polarity. The gas supply of the rifle never runs out. A powerful arc forms between the anode and cathode terminals. Gas molecules and the electrons of the formed arc come into contact. This impact causes the ionization of gas molecules and the release of heat. A heated ionized gas called plasma is fired at the workpiece at a fast rate of speed. The established arc is determined by the rate of gas delivery.

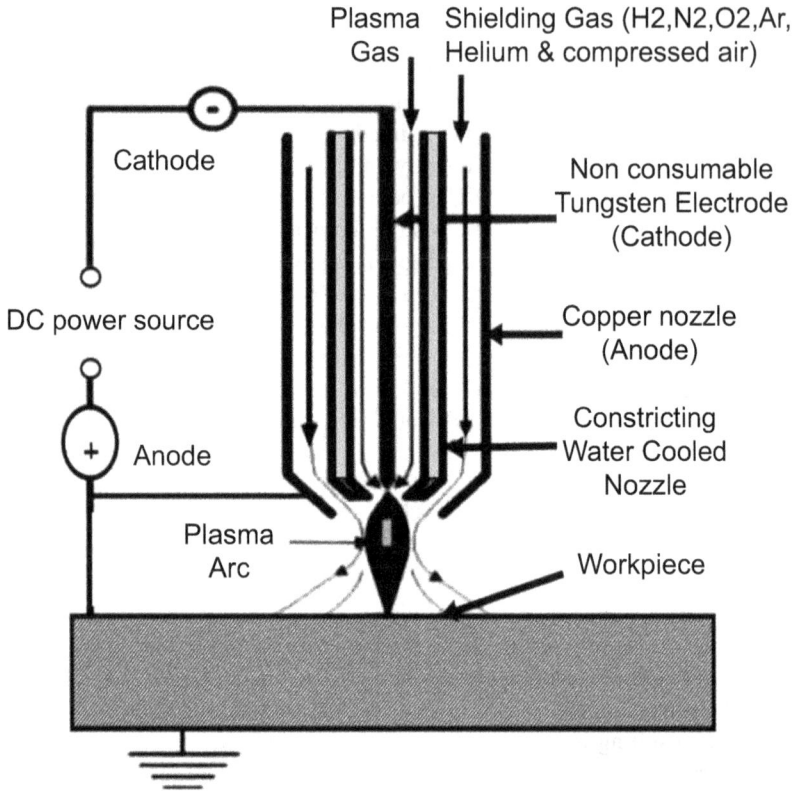

FIGURE 7.6   Schematic diagram of the PAMP.

2. **Power supply:** A power supply (DC) is employed to make two terminals for the plasma gun. The cathode, which is an electrode made of tungsten, is placed within the gun, and the anode, which is the nozzle, is also in the gun. To induce a plasma state of gases, a significant potential difference is applied across the electrodes.
3. **Cooling mechanism:** The plasma gun incorporates a cooling system since the hot gases coming out of the nozzle cause it to heat up. The nozzle is cooled by a water jacket. A water jet surrounds the nozzle.
4. **Tooling:** There is no visible tool in PAM. The cutting tool is a focused spray of hot, plasma-state gases.
5. **Workpiece:** The PAM method can be used to treat a wide range of materials. Aluminum, magnesium, stainless steel, and carbon and alloy steels are among the materials used. The PAM process can be employed on any material that can be processed by the LBMP.

Table 7.6 summarizes the main points of the plasma arc machining process.

TABLE 7.6
Key Points of PAMP

1.	Working principle	**PAMs work is based on the use of ionization plasma to transmit heat, and this high-temperature plasma jet melts the metal and removes it from the workpiece. Gas is forced across an electric arc formed between the cathode and the anode to produce plasma. It works on the premise of melting the contents of the workpiece using a high-temperature plasma jet to remove the material. The material is eliminated using the PAM technique by directing a high-velocity stream of hot gas at high temperature (11,000°C–28,000°C) onto the workpiece.**
2.	Mechanism	The torch generates a high-velocity jet of high-temperature hot gas (plasma) that melts and removes material from the workpiece
3.	Process parameters	Cutting speed, gas pressure, arc current, arc voltage, stand-off distance, and gas flow rate
4.	Components	A power supply, an arc-starting circuit, and a torch
5.	Types of plasma arc	Transferred and non-transferred arc PAM system
6.	Tool electrode	Tungsten
7.	Workpiece material	Stainless steel, mild steel, carbon steel, aluminum, copper, brass, ferrous and non-ferrous alloy
8.	Gas used	Compressed air, nitrogen, argon, hydrogen, and oxygen
9.	MRR	150 cm^3/min
10.	Applications	• Used in mills as well as nuclear submarine pipe systems • Used for welding rocket motor casings • Used in welding stainless steel tubes • Used to cut profiles
11.	Advantages	• This method can easily machine both hard and fragile metals • PAM provides a quicker pace of output • This method can produce small cavities with high-dimensional precision • It is suitable for rough turning of extremely hard materials • It is also employed in machines that repair jet engine blades
12.	Disadvantages	• Plasma arc machining requires expensive equipment • Metallurgical changes occur on the workpiece's surface • Inert gas use is elevated • Shielding is required as oxidation and scale developments occur

7.3.7 ELECTRIC DISCHARGE MACHINING PROCESS

7.3.7.1 Introduction to EDMP

EDM techniques are non-traditional material removal methods with great accuracy and no physical cutting force between tool and workpiece. Material is removed from the workpiece during this process owing to erosion induced by a quickly repeating spark discharge that occurs between the tool electrode and the workpiece in the presence of a dielectric fluid. It is used to mill exceptionally hard and difficult-to-machine materials such as super alloys, tool steels, and tungsten carbides, and to create sharp inner corners, deep holes, complicated geometries, and so forth [20–22,32–34].

7.3.7.2 Working Principle of EDMP

Surplus material is separated from the workpiece during the EDM process owing to wear produced by the rapidly returning electric spark discharge that occurs between the tool electrode and the workpiece. A servomechanism maintains a predetermined spark gap between the workpiece and the tool electrode (Figure 7.7). Both the tool electrode and the workpiece are submerged in a dielectric fluid. The negative terminal is attached to the tool electrode (which acts as a cathode) while the positive terminal is connected to the workpiece (which acts as an anode). When a voltage is provided between the electro-conductive tool electrode and the workpiece material, a sufficient spark forms between the two electrodes. The acceleration of positive ions and electrons forms a discharge channel that becomes electrically conductive. Then, the spark jumps, triggering collisions between electrons and ions to form a plasma channel and an instantaneous decrease in the electrical resistance of the prior channel. With the help of this channel, the current density is able to reach

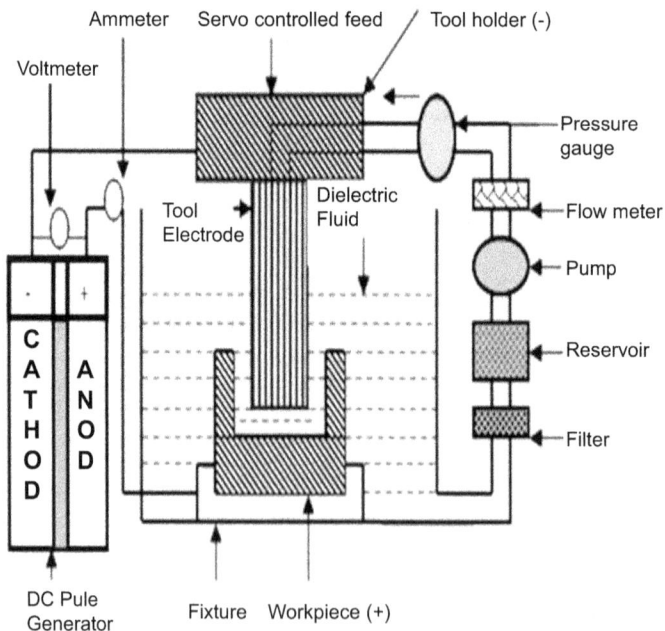

FIGURE 7.7 Line diagram of the electric discharge machining process.

a very high charge, which increases ionization and generates a powerful magnetic field. A small quantity of metal is melted and eroded at such a high pressure and temperature as a result of a sudden spark that develops between the tool electrode and the workpiece and a significant amount of pressure results from the development of these conditions. Due to the quick evaporation and melting of the conductive and hard material, a localized temperature increase of this magnitude results in material decrease [20–22,32–34]. Figure 7.7 depicts the electric discharge machining process.

7.3.7.3 Components of EDM

The main components of the electric discharge machine are briefly described next.

1. **Workpiece:** All electrically conducting materials can be used as workpieces and are machined by a die sinking electric discharge machine.
2. **Tool electrode:** In an EDM process, the electrode is the device that chooses the shape's profile. The qualities of the electrode's material and design should be taken into account when choosing a tool electrode. The main criteria for a tool electrode are that it should be easily accessible, show little tool wear, be electrically conductive, have a good surface quality, and be machinable.
3. **Dielectric fluid:** The electric discharge machine system consists of a reservoir into which the dielectric fluid is filled. Both the tool electrode and workpiece are dipped into the dielectric fluid.
4. **Servo system:** A servo system is used for maintaining the predetermined gap between the tool and the electrode. The servomechanism is commanded by signals from the gap voltage sensor arrangement in the power supply and manages the feed of the tool electrode and workpiece to exactly match the rate of material removal.
5. **Power supply:** A crucial component of the EDM process is the electrical power source. The main utility supply's alternating current (AC) is converted into the direct current pulse required to create the spark discharge at the machining gap by a direct current pulse generator. The pulse generator must be able to regulate electrode polarity, pulse frequency, duty cycle, pulse voltage, and other parameters.

Table 7.7 summarizes the main points of the electric discharge machining process.

7.3.8 ELECTROCHEMICAL MACHINING PROCESS

7.3.8.1 Introduction to ECMP

Electrochemical machining is based on the electrolysis law of Faraday. To eliminate unwanted metal, an electric current is applied. ECM is a process for finishing

TABLE 7.7
Key Points of EDMP

1.	Working principle	The basis of the EDM working process is the production of sparks and the erosion of metal by sparks. EDM spark erosion is when an electric spark strikes a piece of metal and burns a hole into it.
2.	Mechanism	Vaporization
3.	Material removal technique	Using a powerful electric spark
4.	Work material	Electrically conductive metal and alloys
5.	Tool material	Copper alloy, alloy of zinc, copper, and tungsten
6.	Dielectric fluid	Transformer oil, kerosene, paraffin oil, and deionized water
7.	MRR	50–80 mm³/s
8.	Surface finish	• 0.25 µm
9.	Capital cost	Medium
10.	Efficiency	High
11.	Applications	• Production of complicated and irregular-shaped profile • Resharpening of cutting tools • Mold and die making • Prototype manufacturing • Micro-drilling
12.	Advantages	• Cuts extremely hard • Creates complex shapes • Good surface finish can be obtained • Very close tolerances for high precision • No direct contact between tool and workpiece
13.	Disadvantages	• Not suitable for non-conducting material • Slow MRR • Excessive tool wear • Taper effect at the edge of machined cavity

workpiece surfaces using an anodic metal dissolution. The machining tool is the cathode (–), which functions under direct current and in the presence of an electrolyte fluid to produce an anodic reaction that precisely removes workpiece (+) surface material. ECM can efficiently machine materials that are extremely hard. ECM can also produce small or irregularly formed angles, complicated curves, and cavities in extremely hard materials such as titanium, kovar, and carbide. It is now commonly used for machining aircraft components, crucial deburring, fuel injection system components, dies and molds, and other similar applications [19,20,27,29].

7.3.8.2 Working of Electrochemical Machining
The electrochemical machining process begins with the tool advancing toward the workpiece. The tool and the workpiece are separated by a very tiny gap in a suitable electrolyte. When a voltage difference (DC) is applied, the workpiece begins to

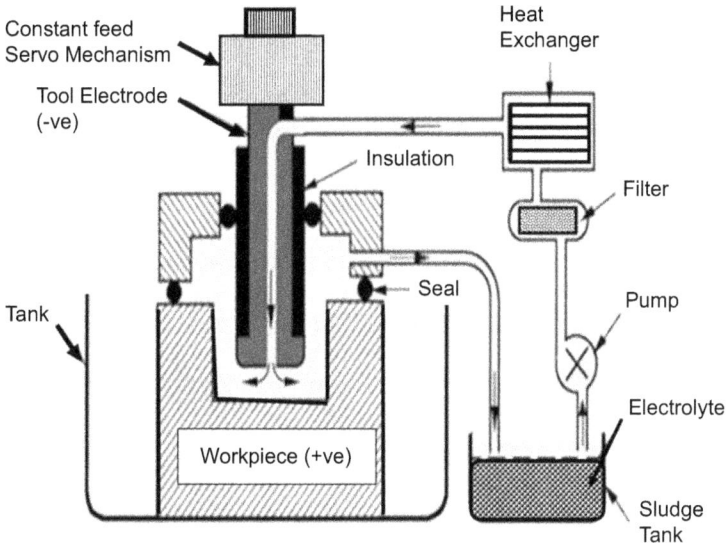

FIGURE 7.8 Schematic diagram of electrochemical machining.

behave as an anode and the tool begins to behave as a cathode. When the electrolysis requirement is met, metal removal from the workpiece begins. The removal is carried out in accordance with the form of the instrument. Because of the flow of electrolytes, material is taken from the workpiece and dropped down in the form of a slug. Then, the electrolyte is filtered by passing it through a centrifuge which removes the slug. The water is then filtered to eliminate any leftover contaminants. If the electrolyte pressure rises, the pressure valve directs the flow of the electrolyte straight to a tank [19, 20, 27, 29]. Figure 7.8 illustrates the electrochemical machining process.

7.3.8.3 Construction of Electrochemical Machining Setup

The following are the different components of an electrochemical plant:

1. **Power supply:** The source of energy used by the setup is the power supply. According to the application, the power source is generally a DC battery with a potential difference of 3–30 V.
2. **Electrolyte:** During the machining process, the workpiece and tool are held in place by an electrolyte, which is a salt solution. It acts as a conductor of the current between the workpiece and the tool. Along with preventing the tool and workpiece from overheating, it also helps remove waste materials from internal gaps and serves as a coolant. The electrochemical method makes use of electrolytes including sodium chloride (NaCl), sodium nitrate (NaNO$_3$), and hydrochloric acid (HCl).
3. **Tool:** Cathode electrodes are the most frequently used electrodes in ECM. Also, it is the best shape to cut the item into. Tools used in ECM should always have accurate dimensions.

TABLE 7.8
Key Points of ECMP

1.	Working principle	ECM is based on the Faraday law of electrolysis, which says that by placing two electrodes in a container filled with conductive liquid or electrolyte and delivering a high ampere DC voltage across them, metal can be plated on the cathode and depleted from the anode (positive terminal) (negative terminal). This is the core idea of electrochemical machining. This machining technique connects the tool to the cathode (negative terminal of the battery) and the workpiece to the anode (positive terminal of the battery) (work as anode). At a shallow depth, they are both immersed in an electrolyte solution. Metal is removed from the workpiece by applying a direct current to the electrode.
2.	Mechanism	Electrolysis (ion dissolution)
3.	Material removal technique	Based on Faraday's law of electrolysis
4.	Work material	Difficult to machine electrically conductive materials
5.	Tool material	Copper tungsten, stainless steel, copper, brass, bronze, and titanium
6.	Medium	Conducting electrolyte (NaCl and $NaNO_3$)
7.	Metal removal rate	27 mm^3/s
8.	Potential volts	10–30 V
9.	Current	10,000 A
10.	Surface finish	0.2–0.8 μm
11.	Gap	0.20 mm
12.	Power	100 kW
13.	Power requirement	Medium
14.	Capital cost	High
15.	Efficiency	Low
16.	Applications	• Used for micromachining because of its excellent precision and surface smoothness • ECM is effective for machining turbine blades due to their intricate concave design • It can also be used for drilling and milling operations • It is utilized to make extremely small gear systems
17.	Advantages	• Negligible stresses • No heat produced • No tool wear • Good surface finish • Accurate dimensions • Compatible with hard materials
18.	Disadvantages	• Higher cost • Hazardous to the environment • Corrosion • Required large area • Limited materials

4. **Mechanical system:** The ECM cannot function without a mechanical system. It is employed for the perpendicular advancement of a tool at a constant speed.
5. **Tank:** A tank is used to store the electrolyte, tool, and workpiece, and it is where all of the reactions take place.
6. **Pressure gauge:** A pressure gauge is used to calculate the amount of electrolytes provided to the tool.
7. **Flow control valve:** The volume of electrolytes that flow to the tool is controlled by a flow control valve.
8. **Pressure relief valve:** If the electrolyte flow pressure exceeds a specific threshold, the pressure relief valve opens and returns the electrolyte to the tank.
9. **Reservoir tank:** The storage tank where pure electrolytes are kept.
10. **Filter and centrifuge:** Prior to the electrolyte entering the reservoir tank, a filter is used to clean it. It stops the accumulation of excess electrolytes. The electrolyte and slug are separated by a centrifuge.
11. Slug container: The electrolyte-free slug is kept separate in a slug container. Many experiments can be performed using this slug.

Table 7.8 summarizes the key points of the ECMP.

ACKNOWLEDGMENTS

Dr R.S. Jadoun and Dr Sushil Kumar Choudhary would like to thank G.B. Pant University of Agriculture & Technology, Pantnagar, U.S. Nagar, Uttarakhand, India, and the Department of Mechanical and Automobile Engineering, University Institute of Engineering & Technology, Babasaheb Bhimrao Ambedkar University (Central University), Lucknow, India, for their cooperation and contributions. They are also grateful to the reviewers, editorial board, and the publication and marketing teams for their valuable suggestions and ideas for further improving the quality of the current work.

REFERENCES

1. Sawhney GS. *Manufacturing Science-II*. Chennai: Scitech Publication (India) Pvt. Ltd; 2010.
2. Kalpakyian S and Schmid SR. *Manufacturing Engineering and Technology*. 4th edition. New York: Prentice-Hall; 2001.
3. Upadhyay V and Agarwal V. *Basic Manufacturing Process*. New Delhi: S.K. Kataria & Sons, Publisher of Engineering and Computer Books; 2014.
4. Schey J. *Introduction to Manufacturing Processes*. New York: McGraw-Hill; 2000.
5. Sandy K. *Manufacturing Process-II*. New Delhi: S.K. Kataria & Sons, Publisher of Engineering and Computer Books; 2013.
6. Raghuwanshi BS. *Workshop Technology Vol. I & II*. Delhi: Dhanpath Rai & Sons; 2003.
7. Degarmo EP, Black JT and Kohser RA. *Materials and Processes in Manufacturing*. 8th edition. New York: Wiley; 1997.

8. Benedict GF. *Non- Traditional Manufacturing Process*. New York: Marcel Dekker Inc; 1987.
9. Bhattacharya A. *New Technology*. Calcutta: The institute of Engineers (I); 1973.
10. Benedict GF. Nontraditional manufacturing processes. *Manufacturing Engineering and Materials Processing*. New York: Marcel Dekker Inc., 1987.
11. John KC. *Mechanical Workshop Practice*. 2nd edition. New Delhi: PHI; 2010.
12. Kannaiah P and Narayana KL. *Workshop Manual*. 2nd edition. Chennai: Scitech Publishers; 2009.
13. Zenger DC and Boothroyd G. Selection of manufacturing processes and materials for component parts. In *Proceedings 4th International Conference on Product Design for Manufacture and Assembly*, Rhode Island, 1989.
14. Anup G and Jacob Moses A. *Unconventional Machining Process*. Pune: Technical Publications; 2020.
15. Jain V, Sharma AK and Kumar P. Recent developments and research issues in micro-ultrasonic machining. *ISRN Mechanical Engineering*, 2011; 413231.
16. Goetze D. Effect of vibration amplitude, frequency, and composition of the abrasive slurry on the rate of ultrasonic machining in ketos tool steel. *The Journal of the Acoustical Society of America*, 1956; 28(6):1033–1037.
17. Hu P, Zhang JM, Pei ZJ and Treadwell C. Modeling of material removal rate in rotary ultrasonic machining: Designed experiments. *Journal of Materials Processing Technology*, 2002; 129(1–3):339–344.
18. Kennedy DC and Grieve RJ. Ultrasonic machining-a review. *Production Engineering*, 1975; 54(9):481–486.
19. Mc Geough JA. *Advance Methods of Machining*. London: Chapman & Hall; 1988.
20. Mishra PK. *Non Conventional Machining*. New Delhi: Narosa Publishing House; 2006.
21. Pandey PC and Shan HS. *Modern Machining Processes*. New Delhi: Tata McGraw Hill Publishing Co. Ltd; 1980.
22. Groover MP. *Fundamentals of Modern Manufacturing*. 2nd edition. New York: Wiley; 2002.
23. Sarkar PK and Pandey PC. Some investigations in abrasive jet machining. *Journal of The Institution of Engineers (I)*, 1976; 284:56.
24. Verma AP and Lal GK. An experimental study of abrasive jet machining. *International Journal of Machine Tool Design and Research*, 1984; 1(24):19–29.
25. Verma AP and Lal GK. Basic mechanics of abrasive jet machining. *Journal of Industrial and Production Engineering*, 1985; 66:74–81.
26. Ingula CN. Abrasive jet machining. *Tool and Manufacturing Engineers*, 1967; 59:28–36.
27. Jagadeesh T. *Unconventional Machining Processes*. Delhi-India: I. K. International Publishing House Pvt. Ltd.; 2016.
28. Dubey AK and Yadava V. Laser Beam Machining – A review. *International Journal of Machine Tools and Manufacture*, 2008; 48(6):609–628.
29. Von Gutfeld RJ and Sheppard KG. Electrochemical micro-fabrication by laser-enhanced photothermal processes. *IBM Journal of Research and Development*, 1998; 42(5):639–653. https://doi.org/10.1147/rd.425.0639.
30. Von Gutfeld RJ, Tynan EE, Melcher RL and Blum SE. Laser enhanced electroplating and maskless pattern generation. *Applied Physics Letters*, 1979; 35(9):651–653. https://doi.org/10.1063/1.91242.
31. Senthilkumar V. Laser cutting process – A review. *International Journal of Darshan Institute on Engineering Research & Emerging Technologies*, 2014; 3:44–48.

32. Choudhary SK, Jadoun RS and Tomar P. Optimization of EDM process parameters for TWR on machining of inconel 600 superalloy using Taguchi approach. *Journal of Materials Today: Proceedings*, 2022; 5(57):2281–2288.

33. Jain VK. Analysis of Electro discharge drilling of a blind hole in HSS using Bit type of tools. *Microtechnic*, 1989; 2:34.

34. Choudhary SK and Jadoun RS. Optimization of EDM process parameters for MRR of inconel 600 using Taguchi method. *International Journal of Mechanical and Production Engineering Research and Development (IJMPERD)*, 2020; 4(10):11481–11492.

8 Design and Development of Conventional Processes for Medical Implants

Nishant K. Singh, Rajeev K. Upadhyay,
Virendar Kumar, and Abhishek Sharma

8.1 INTRODUCTION

Medical implants play a crucial role in improving the quality of life for patients with various medical conditions. The design and development of these implants are essential to ensure their safety and efficacy. Medical implants are devices or materials that are placed inside the body to replace a missing body part, improve bodily function, or treat a medical condition. Examples include artificial joints, heart pacemakers, insulin pumps, cochlear implants, and spinal cord stimulators [1]. Medical implants are designed to be long-lasting and biocompatible, meaning they do not cause harm to the surrounding tissue. They can be made from materials such as metals, ceramics, or biodegradable polymers. They are surgically placed and may require follow-up procedures or replacement over time. Risks associated with medical implants can include infection, device failure, and rejection by the body's immune system [2]. This chapter provides an overview of the conventional processes involved in the design and development of medical implants.

8.2 DESIGN PROCESS

The design process of medical implants begins with a thorough understanding of the medical condition that the implant is to address. This involves reviewing existing medical literature, talking to healthcare professionals, and understanding the patient's needs. The next step is to design the implant based on these requirements, taking into consideration factors such as the implant's size, shape, and material. The design must be safe and effective, and must also meet regulatory requirements [3]. Medical implants can be designed using a variety of design processes, depending on the type of implant and the goals of the design. The following are some of the most commonly used design processes for medical implants (Figure 8.1).

DOI: 10.1201/9781003375098-8

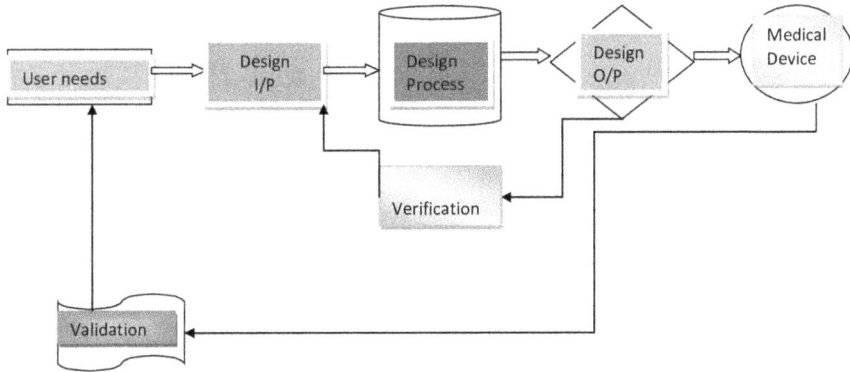

FIGURE 8.1 Methodology used in medical implant design.

8.2.1 BIO-INSPIRED DESIGN

The bio-inspired design process involves drawing inspiration from nature to create medical implants that mimic the structure and function of natural tissues. For example, the design of a bio-inspired artificial heart valve might be based on the anatomy and mechanics of a healthy heart valve [4].

8.2.2 COMPUTER-AIDED DESIGN (CAD)

CAD is a digital design process that allows designers to create detailed 3D models of medical implants using computer software. This process allows for precise measurements, simulations, and visualization of the implant before it is manufactured [5].

8.2.3 RAPID PROTOTYPING

Rapid prototyping is a process that allows designers to quickly create prototypes of medical implants using 3D printing or other similar technologies. This allows designers to test and refine the design before manufacturing the final product [6].

8.2.4 FINITE ELEMENT ANALYSIS (FEA)

FEA is a numerical simulation process that allows designers to predict how a medical implant will behave under different loads and conditions. This process helps to identify potential design flaws and optimize the implant for performance and safety [7].

8.2.5 HUMAN FACTORS ENGINEERING

The human factors engineering design process takes into account the anatomy, physiology, and ergonomics of the human body when designing medical implants. The goal is to create implants that are comfortable, easy to use, and functional for the patient [8].

These are some of the most commonly used design processes for medical implants, but there may be others that are specific to certain types of implants or design goals. The design process used will depend on factors such as the implant's intended use, the materials being used, and the available technology.

8.3   DEVELOPMENT PROCESS

Once the design is complete, the development process begins. This involves the creation of prototypes, testing and evaluation, and refinement of the design. The prototypes are tested for their safety and efficacy, and any issues identified are addressed. The implant's design is also optimized to ensure that it is easy to use and cost-effective. Medical implants are devices designed to be placed inside the body for therapeutic or diagnostic purposes. The development of these devices is a complex process that involves multiple stages and requires a deep understanding of both the medical and engineering aspects. Different approaches to the development process are used depending on the type of implant, its intended use, and the needs of the target population. This chapter discusses some of the most common development processes used in the design and production of medical implants.

8.3.1   TRADITIONAL DESIGN AND DEVELOPMENT PROCESS

The traditional design and development process is the most common approach used in medical implant development. It involves a series of sequential stages, starting with the conceptualization and design of the implant, followed by prototyping, testing, and finally manufacturing. This process is often time-consuming and can result in multiple iterations of the design, but it is well established and has proven effective for many types of implants [9].

8.3.2   AGILE DEVELOPMENT PROCESS

A newer approach is the Agile development process which is becoming increasingly popular in medical implant development. It is based on the Agile methodology, which emphasizes collaboration and adaptability, and is designed to be more responsive to changing needs and market conditions. This process involves rapid iteration and continuous testing and improvement, allowing teams to quickly make changes to the design based on feedback from stakeholders and users [10].

8.3.3   RAPID PROTOTYPING PROCESS

The rapid prototyping process involves using rapid prototyping techniques, such as 3D printing, to quickly create the physical models of an implant for testing and evaluation. This allows engineers and physicians to quickly identify and address any design or performance issues, reducing the time to market for the final product [11].

8.3.4   Human-Centered Design Process

The human-centered design process focuses on understanding and incorporating the needs and preferences of the target population into the design of an implant. It involves close collaboration with patients, physicians, and other stakeholders to gather information and feedback on the design, and to ensure that the implant meets the needs and expectations of the target population [12].

> In conclusion, the development process used for medical implants depends on the specific needs of the implant and the target population. Whether using a traditional, Agile, rapid prototyping, or human-centered design process, it is important to carefully consider all factors that may impact the design, performance, and safety of the final product.

8.4   MANUFACTURING PROCESS

Once the design is finalized, the implant is manufactured. This process involves the use of various manufacturing techniques such as molding, casting, and machining. The manufacturing process must be closely monitored to ensure that the implant is of a high quality and meets regulatory requirements. Medical implants are devices used to replace or support biological structures or functions. Several manufacturing processes are used for the production of medical implants, each with its own advantages and disadvantages. The following is an overview of some of the most commonly used processes.

8.4.1   Additive Manufacturing (AM) or 3D Printing

The additive manufacturing or 3D printing process involves building up a product layer by layer using materials such as plastics, metals, and ceramics. The advantages of AM include the ability to produce complex geometries, lower material waste, and reduced production time. Examples of medical implants produced using AM include spinal implants, dental implants, and customized prosthetics [13].

8.4.2   CNC Machining

The CNC machining process uses computer-controlled cutting tools to shape a solid block of material into the desired implant design. The advantages of CNC machining include precise control of dimensional accuracy, a wide range of compatible materials, and efficient production times. Examples of medical implants produced using CNC machining include hip and knee implants, spinal fusion devices, and dental implants.

8.4.3   Injection Molding

The process of injection molding involves injecting melted material into a mold to produce a part with the desired shape and size. The advantages of injection molding include low production costs, efficient production times, and the ability to produce parts with complex geometries. Examples of medical implants produced using injection molding include catheter tubing, pacemaker housing, and drug delivery devices [14].

8.4.4 FORGING

The forging process involves heating a material to its plastic state and then using high pressure to shape it into the desired form. The advantages of forging include improved strength and toughness compared to cast parts, and the ability to produce parts with consistent mechanical properties. Examples of medical implants produced using forging include spinal fusion devices and joint replacement implants.

8.4.5 CASTING

The casting process involves pouring melted material into a mold to produce a solid part with the desired shape. The advantages of casting include the ability to produce parts with complex geometries, lower production costs, and the ability to use a wide range of materials. Examples of medical implants produced using casting include dental implants and orthopedic implants [15].

8.5 REGULATORY APPROVAL

The final step in the development of medical implants is obtaining regulatory approval. This involves submitting the implant to regulatory agencies such as the US Food and Drug Administration (FDA) for approval. The regulatory approval process ensures that the implant is safe and effective for use in patients. Regulatory approval is a crucial step in the development of medical implants, as it ensures that the product meets the necessary standards for safety, effectiveness, and quality. The process of obtaining regulatory approval for medical implants typically involves the following steps.

8.5.1 PRECLINICAL TESTING

Before a medical implant can be tested in humans, it must undergo extensive preclinical testing. This usually includes laboratory tests, animal testing, and simulations. The aim of preclinical testing is to evaluate the safety and performance of the implant, and to identify any potential issues that need to be addressed before clinical trials.

8.5.2 CLINICAL TRIALS

After preclinical testing is completed, the next step is to conduct clinical trials. These trials are conducted on human subjects and are designed to evaluate the safety and effectiveness of the implant in a real-world setting. Clinical trials typically involve three phases: Phase I trials, which are small trials designed to evaluate the safety of the implant; Phase II trials, which are larger trials that evaluate the effectiveness of the implant; and Phase III trials, which are large, multi-center trials that provide further evidence of the implant's safety and effectiveness.

8.5.3 REGULATORY FILING

After successful completion of clinical trials, the next step is to file for regulatory approval. The specific regulatory agency that approves medical implants varies by country, but in the United States, the FDA's Center for Devices and Radiological Health (CDRH) is responsible for reviewing and approving medical implants. In order to obtain approval, the manufacturer must submit a pre-market approval (PMA) application, which includes data from preclinical testing and clinical trials, as well as information about the implant's design, manufacturing processes, and labeling.

8.5.4 REVIEW AND APPROVAL

After a PMA application is submitted, the regulatory agency will review the data and make a decision about whether to approve the implant. This process can take several months to a year or more, depending on the complexity of the product and the agency's review schedule. If the implant is approved, it can be commercially marketed and used by healthcare providers. The process of obtaining regulatory approval for medical implants is complex and time-consuming, but it is necessary to ensure the safety and effectiveness of these products for patients.

In conclusion, the design and development of medical implants involve a comprehensive and multi-step process that includes design, development, manufacturing, and regulatory approval. This process ensures that medical implants are safe, effective, and meet regulatory requirements.

8.6 RECENT DEVELOPMENT IN THE FIELD OF MEDICAL IMPLANTS

8.6.1 RECENT ADVANCES IN DESIGN PROCESS ADOPTED IN THE FIELD OF MEDICAL IMPLANTS

In recent years, the design process for medical implants has shifted toward a more patient-centric approach, incorporating the use of advanced technologies such as computer-aided design, computer-aided manufacturing (CAM), and 3D printing. This has allowed for a more precise and customizable design, as well as faster and more efficient production. One example of this is the development of personalized spinal implants, which use patient-specific imaging data and 3D printing technology to create implants that are tailored to the individual's anatomy. This results in a better fit and improved patient outcomes [16]. Another development is the use of biocompatible materials in the design of medical implants. Materials such as titanium and ceramics minimize the risk of adverse reactions and improve the longevity of the implant [17]. The use of simulation and testing techniques, such as finite element analysis and in vitro testing, has also become more prevalent in the design process for medical implants. These techniques allow for the evaluation of implant performance under various conditions and help to identify potential design issues before the implant is used in a clinical setting [18]. Overall, the advancements in technology and design processes for medical implants have led to improved patient outcomes, increased safety, and reduced costs.

8.6.2 RECENT ADVANCES IN DEVELOPMENT PROCESS ADOPTED IN THE FIELD OF MEDICAL IMPLANTS

Medical implants are devices used to replace or augment damaged or missing biological tissues. The development process of medical implants has seen significant advancements in recent years. The following are some of the key areas of focus (Figure 8.2).

8.6.2.1 Biomaterials

There has been an increased focus on developing biocompatible and biodegradable materials that can mimic the properties of natural tissues. These materials include polymers, ceramics, and metals such as titanium and magnesium [19]. Different kinds of biocompatible materials are depicted in Figure 8.2.

8.6.2.2 Surface Modification

The surface of medical implants is critical to their performance, and recent advances in surface modification techniques have improved the biocompatibility of implants. This includes techniques such as plasma surface modification and electrospinning.

8.6.2.3 Manufacturing Techniques

Additive manufacturing techniques, such as 3D printing, have been developed for the production of medical implants. These techniques allow for the creation of complex and customized implant designs, leading to improved patient outcomes [20].

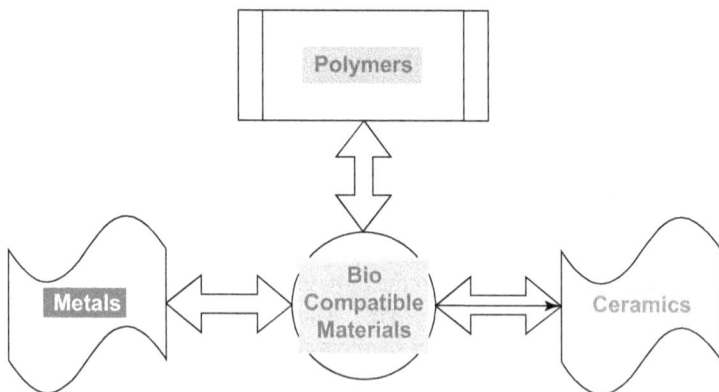

FIGURE 8.2 Biocompatible materials used in medical implants.

8.6.2.4 Smart Implants

The development of smart medical implants has been a major area of focus in recent years. These implants have the ability to monitor and respond to changes in the body, and can provide real-time information to healthcare providers [21].

8.6.2.5 Drug Delivery Systems

There has been significant research into the development of medical implants that can deliver drugs directly to the site of disease or injury. This has the potential to improve treatment outcomes and reduce the side effects associated with systemic drug administration [22].

8.6.3 RECENT ADVANCES IN MANUFACTURING PROCESS ADOPTED IN THE FIELD OF MEDICAL IMPLANTS

Advances in manufacturing processes for medical implants have enabled improved designs, increased reliability, and reduced production costs.

8.6.3.1 Additive Manufacturing (3D Printing)

Additive manufacturing, also known as 3D printing, has revolutionized the way medical implants are manufactured. It allows for the creation of complex, custom-fit implants with improved mechanical properties, reduced waste and cost, and improved production times compared to traditional manufacturing processes [23]. Figure 8.3 illustrates the steps involved in AM/3D printing.

8.6.3.2 Advanced Materials

The use of advanced materials such as titanium, cobalt-chrome alloys, and ultra-high molecular weight polyethylene (UHMWPE) has also improved the functionality and durability of medical implants. For example, UHMWPE has been widely adopted as a material for orthopedic implants due to its high wear resistance and biocompatibility [24].

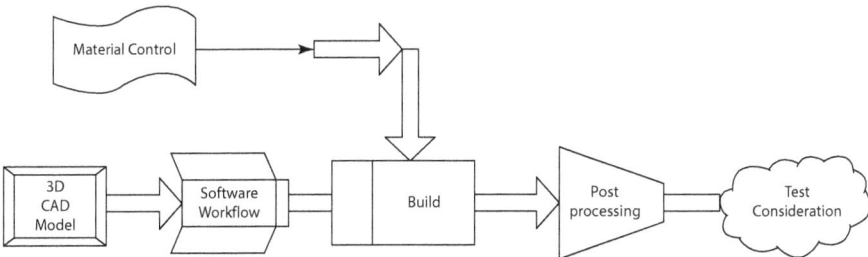

FIGURE 8.3 Steps involved in AM/3D printing.

8.6.3.3 Surface Modification

Surface modification techniques, such as plasma spray, ion implantation, and electrospinning, have been developed to improve the biocompatibility and performance of medical implants. These techniques allow for the creation of rough or porous surfaces that promote tissue ingrowth, improve implant stability, and reduce the risk of infection [25].

8.7 EXISTING PROBLEMS IN MEDICAL IMPLANTS

Medical implants have come a long way in improving the quality of life for many people; however, several problems exist with these devices that need to be addressed.

- **Rejection and infection:** One of the biggest problems with medical implants is the risk of rejection or infection. The body's immune system may recognize the implant as a foreign object and attempt to reject it, causing inflammation and potentially damaging the implant or surrounding tissue. Additionally, infections can occur at the site of implantation, which can also cause inflammation and potentially lead to implant failure [26].
- **Implant failure:** Another issue with medical implants is the potential for failure. Some devices can malfunction or break down over time, requiring replacement or revision surgery. In some cases, implants can even cause further harm to the patient, such as internal damage to surrounding organs or tissues [27].
- **Limited lifespan:** Many medical implants have a limited lifespan and need to be replaced after a certain period of time. This can be a major inconvenience for patients, who must undergo additional surgeries and recovery periods to have the device replaced [28].
- **Complicated implantation:** The implantation of some medical devices can be a complex and challenging procedure, requiring specialized training and experience. In some cases, complications during the implantation process can lead to further problems with the device or even harm to the patient [29].
- **Cost:** Medical implants can be expensive, and many people are unable to afford the cost of these devices or the surgeries required to implant them. This can lead to unequal access to healthcare for those in need, particularly for those who do not have adequate insurance coverage [30].

These are just some of the existing problems with medical implants, but it is important to note that significant progress has been made in addressing these issues and improving the safety and efficacy of these devices.

8.8 MARKET ANALYSIS OF MEDICAL IMPLANTS

The global medical implant market is a rapidly growing industry, driven by advancements in technology, an increasing aging population, and the growing prevalence of chronic diseases. Medical implants are devices that are surgically implanted into the body to replace or support damaged or missing biological structures. Some common examples of medical implants include artificial joints, pacemakers, implantable cardioverter defibrillators (ICDs), and dental implants.

According to a report by Allied Market Research, the global medical implant market was valued at $67.8 billion in 2016 and is expected to reach $133.9 billion by 2023, growing at a compound annual growth rate (CAGR) of 9.4% from 2017 to 2023. The report cites the increasing prevalence of chronic diseases, such as osteoarthritis, as a major factor driving the growth of the market. In terms of geography, North America is the largest market for medical implants, followed by Europe and Asia-Pacific. The North American market is driven by factors such as a high prevalence of chronic diseases, an aging population, and the presence of an advanced healthcare infrastructure and reimbursement policies. The Asia-Pacific region is expected to experience the highest growth rate during the forecast period, driven by factors such as increasing access to healthcare and an expanding middle class with increasing disposable income. The medical implant market is highly competitive, with a large number of players operating in the industry. Some of the leading players in the market include Johnson & Johnson, Stryker Corporation, Zimmer Biomet Holdings, Inc., Smith & Nephew, and Medtronic. These companies are investing heavily in research and development to bring new and advanced products to the market.

In conclusion, the global medical implant market is expected to experience significant growth in the coming years, driven by advancements in technology, an aging population, and the growing prevalence of chronic diseases. North America and Europe are the largest markets for medical implants, while the Asia-Pacific region is expected to experience the highest growth rate during the forecast period. The market is highly competitive, with a large number of players operating in the industry.

8.9 FUTURE RESEARCH DIRECTION IN THE DESIGN AND DEVELOPMENT OF MEDICAL IMPLANTS

In the future, the design and development of medical implants are likely to focus on several key areas:

- **Personalized medicine:** The development of medical implants that are personalized to the specific needs of an individual, taking into account factors such as genetics, lifestyle, and environment.
- **Biocompatibility:** The use of biocompatible materials in medical implants that reduce the risk of rejection and inflammation, as well as improve integration with the body.
- **Wireless communication:** The integration of wireless communication technology into medical implants to enable real-time monitoring and control, as well as improve patient outcomes.
- **Artificial intelligence:** The integration of artificial intelligence into medical implants to enable more accurate diagnosis and treatment, as well as improve patient outcomes.

- **Energy harvesting:** The development of medical implants that are self-powered, either through energy harvesting technologies such as piezoelectricity or through the use of long-lasting batteries.
- **Smart implants:** The development of medical implants that are equipped with sensors and other intelligent features, enabling real-time monitoring and feedback for improved patient outcomes.
- **Minimally invasive surgery:** The development of medical implants that can be implanted using minimally invasive surgical techniques, reducing the risk of complications and improving patient outcomes.

Overall, the future direction of medical implant design and development is focused on improving patient outcomes, increasing patient safety, and reducing the risk of complications associated with implantation.

REFERENCES

1. FDA, Medical devices: Overview. 2021. Retrieved from https://www.fda.gov/medical-devices/overview.
2. Medical Device Development Process. 2021. Retrieved from https://www.medicaldesignbriefs.com/design-process/medical-device-development-process.
3. A Guide to the Medical Device Design Process. 2021. Retrieved from https://www.qmed.com/feature/guide-medical-device-design-process.
4. S. Haddadi, S. Khanmohammadi, and M. Shakeri, "A review on bio-inspired design of artificial heart valves," *World Journal of Engineering and Technology*, 6(4), pp. 362–370, 2018.
5. J. van den Dolder, R. van Limbeek, J. C. M. M. de Lange, and E. F. W. de Boer, "Computer-aided design in orthopaedic implant design," *Journal of Orthopaedic Surgery and Research*, 8(1), pp. 1–9, 2013.
6. C. J. Pecho, S. M. Hsu, S. M. Kuo, T. W. Chan, and Y. H. Lee, "The application of rapid prototyping in orthopaedic implant design," *Journal of Medical Systems*, 36(2), pp. 579–586, 2012.
7. T. C. Lu, Y. H. Lee, and W. J. Chen, "Finite element analysis in orthopaedic implant design: A review," *Journal of Medical Systems*, 38(4), pp. 2047–2058, 2014.
8. K. L. McGrath and L. A. Stevens, "Human factors engineering in medical device design," *Annual Review of Biomedical Engineering*, 11, pp. 175–199, 2009.
9. "Medical device design and development: An overview," *Medical Device Academy*, 2020.
10. "Agile medical device development: Benefits and best practices," *HCL Technologies*, 2021.
11. "Human-centered design in medical device development," *Design World*, 2021.
12. "Rapid prototyping for medical devices," *Protolabs*, 2021.
13. D. M. Anderson and A. Atala, "Additive manufacturing for medical applications," *Journal of Biomedical Materials Research. Part B, Applied Biomaterials*, 101(7), pp. 1637–1649, 2013.
14. J. J. Bourke, "Orthopaedic implant manufacturing," *Journal of Medical Engineering and Technology*, 34(2), pp. 92–100, 2010.
15. L. Guan, J. Zhang, and B. Yu, "Advancements in medical implants: Materials, design, and manufacturing," *Journal of Medical Systems*, 44(7), p. 422, 2020.

16. Personalized spinal implants: https://www.ncbi.nlm.nih.gov/pmc/articles/PMC6349193/.
17. Biocompatible materials in medical implants: https://www.sciencedirect.com/science/article/pii/S1359645420307126.
18. Simulation and testing in medical implant design: https://www.ncbi.nlm.nih.gov/pmc/articles/PMC7349862/.
19. S. Sutradhar, S. Das, and J. Bhowmick, "Biodegradable polymers as implant materials," *Progress in Polymer Science*, 93, pp. 1–24, 2019.
20. Y. Lu, Y. Guo, and Y. Wang, "Additive manufacturing of medical implants: Materials, processing, and challenges," *Advanced Engineering Materials*, 21(2), p. 2000073, 2019.
21. H. S. Kim, K. Lee, and S. Kim, "Smart medical implants: Current status and future perspectives," *Biosensors and Bioelectronics*, 153, pp. 1115–1127, 2020.
22. D. D. O'Connor, T. W. Paterson, and A. J. Teh, "Drug delivery systems for medical implants," *Advanced Drug Delivery Reviews*, 145, pp. 1–16, 2019.
23. D. Munoz-Espin and L. E. Murr, "Additive manufacturing of medical implants," *Progress in Materials Science*, 97, pp. 361–410, 2018.
24. N. Cho, K. Song, and J. W. Rhim, "Advanced materials for medical implants," *Advanced Healthcare Materials*, 8(20), p. 1900558, 2019.
25. R. Saravanan and K. Palanivelu, "Surface modification techniques for improving the biocompatibility of medical implants," *Journal of Biomedical Materials Research. Part B: Applied Biomaterials*, 105(5), pp. 1055–1066, 2017.
26. https://www.ncbi.nlm.nih.gov/pmc/articles/PMC6075273/.
27. https://www.ncbi.nlm.nih.gov/books/NBK470333/.
28. https://www.ncbi.nlm.nih.gov/pmc/articles/PMC6111798/.
29. https://www.ncbi.nlm.nih.gov/pmc/articles/PMC5624824/.
30. https://www.ncbi.nlm.nih.gov/pmc/articles/PMC6228039/.

9 Additive Manufacturing in Biomedical Applications
Process, Modeling, and Optimization

*Mohammad Taufik, Vishal Francis,
and Ankit Nayak*

9.1 INTRODUCTION

Additive manufacturing (AM) is a layer-by-layer manufacturing approach to fabricate very complex and intricate features with ease over a short period of time. The process is very useful for biomedical applications, fabricating prosthetics, tailored patient-specific implants, as well as numerous surgical equipment and medical devices. AM has completely changed the way medical applications are used [1,2]. The key function of AM is that really difficult components can be produced in a single setup which would otherwise be impossible to manufacture using any other method. Now, any complex anatomical feature can be printed in a matter of hours by manipulating medical scans. Many healthcare and medicine applications are specific to the individual patient. Prosthetics and implants, for instance, are made specifically for each customer and vary depending on the individual. As a result, these items cannot be mass manufactured. On the other hand, AM is capable of fabricating items in a shorter time span, with good accuracy and optimized parameters. Therefore, the AM process is widely used for various biomedical applications [3–5].

Figure 9.1 shows some of the AM applications in the biomedical sector. The applications area is not limited and more research is being directed to explore various other applications such as the bioprinting of organs.

9.2 ADDITIVE MANUFACTURING PROCESSES FOR BIOMEDICAL APPLICATIONS

Additive manufacturing is now becoming popular in the medical domain due to its ability to fabricate complex geometry parts. AM is used to manufacture implants

DOI: 10.1201/9781003375098-9

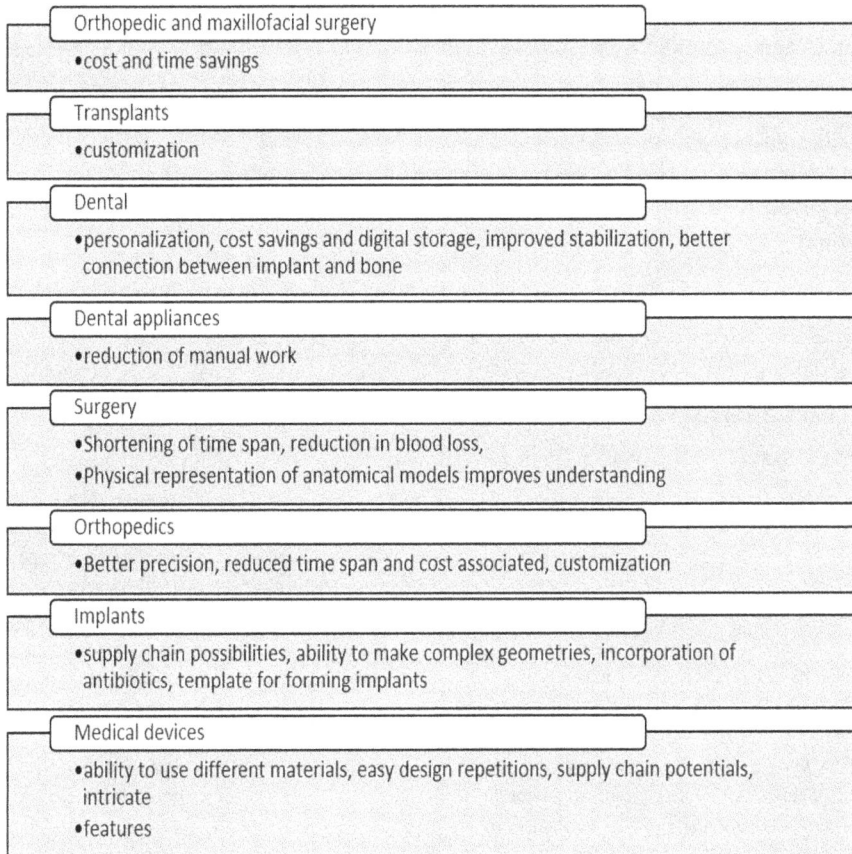

FIGURE 9.1 Some applications of AM in the biomedical sector and their advantages [7].

and surgical guides including endodontic guides for efficient root canal shaping. On the other hand, the application of rapid prototyping in medicine is a widespread practice to explain and understand anatomical details. Parts printed with rapid prototyping are used for surgical planning and classroom demonstrations.

AM consists of preprocessing, processing, and post-processing. In preprocessing, the model is designed using different computer-aided design (CAD) software. Geometries that consist of primitive shapes can be easily designed by editing the shapes with Boolean operations. However, for complicated designs such as sculptures and human body parts, point cloud processing is used to generate a stereolithography (STL) model. The 3D scanning method can be used to copy the external design of any complex shape. 3D scanning provides the point cloud data that can be converted into STL files followed by 3D printing.

The AM process can be divided into seven categories as per the American Society for Testing and Materials (ASTM) standards [6]: directed energy deposition, binder jetting, material jetting, sheet lamination, vat polymerization, powder bed fusion, and material extrusion. Table 9.1 illustrates the classification of various AM processes along with the form of material they use and other common names by which these processes are known.

Even though each process is different based on the principle involved, the basic steps used in the AM process are common, such as CAD modeling, slicing, and printing. Figure 9.2 illustrates the basic steps involved in the AM process. It can be seen that most of these steps remain unchanged irrespective of the process.

TABLE 9.1
Classification of Various AM Processes [7]

AM Category	Working principle	State of Material	AM processes
• Powder bed fusion (PBF)	• thermal energy fuses regions of a powder bed	• powder	• selective laser sintering (SLS), direct metal laser sintering (DMLS), selective laser melting (SLM)
• Material extrusion (MEX)	• material dispensed through a nozzle	• filament, • pellets, paste	• fused deposition modeling (FDM), (fused filament fabrication) FFF
• VAT photo-polymerization	• liquid photopolymer in a vat is cured by light	• Liquid	• SLA, digital light projection (DIP)
• Material jetting (MJ)	• droplets of material are selectively deposited	• Liquid	• PolyJet, NJP
• Binder jetting (BJ)	• a liquid bonding agent is selectively deposited	• Powder	• 3D printing (3DP), Color Jet printing (CJP)
• Sheet lamination (SL)	• sheets of material are bonded	• Sheets	• laminated object manufacturing (LOM), ultrasonic additive manufacturing (UAM)
• Directed energy deposition (DED)	• focused thermal energy used to fuse materials by melting when depositing	• powder, wire	• laser-engineered net shaping (LENS), EBAM

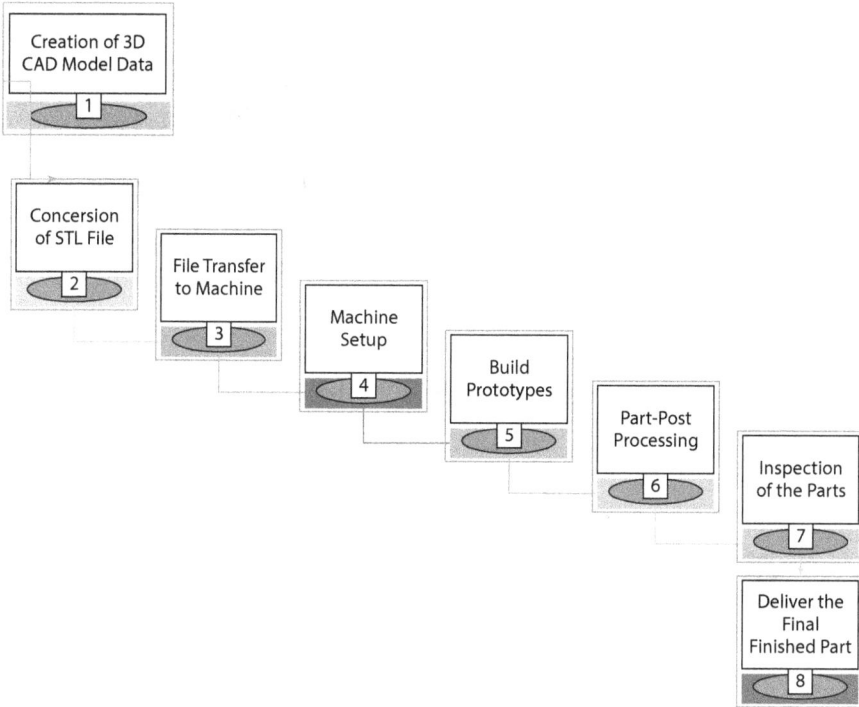

FIGURE 9.2 Basic steps of the AM process [13].

AM is utilized for various biomedical applications such as prostheses, implants, models, and tools [7]. A major reason for this utilization is the advantages achieved by using AM in the biomedical sector. The advantages include time-saving, improvements in mechanical properties, fabrication of complex geometries, lightweight parts, presurgical planning, and customized features and parts.

Various technologies of the AM process utilized for numerous biomedical applications are illustrated in Figures 9.3–9.6. Fused deposition modeling (FDM) is an extrusion-based process in which the material in the form of a filament is heated and extruded through a nozzle in a controlled manner (Figure 9.3). Polymers can be used in this process. Compared to other AM processes, this process is simpler and less post-processing is required. However, the process also has certain limitations in terms of surface finish and resolution. The FDM process can be utilized for the fabrication of scaffolds for tissue engineering, presurgical planning, medical devices, and prostheses.

The range of applications is restricted based on the materials that each AM method can process.

Another AM process that is utilized for biomedical applications is selective laser sintering (SLS). Figure 9.4 is a schematic diagram of an SLS system. The process is additive in nature and produces models and prototype parts from 3D CAD models,

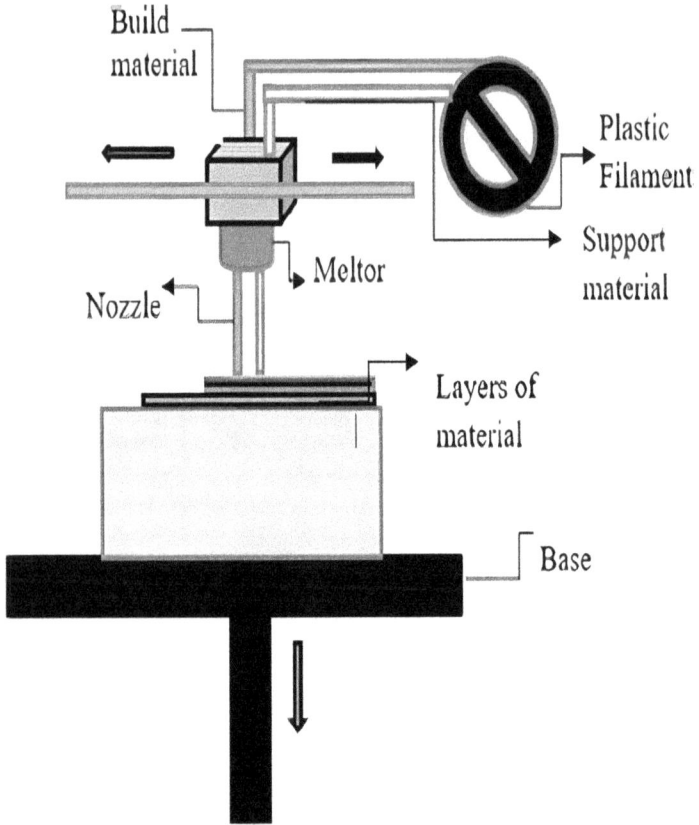

FIGURE 9.3 Schematic of the FDM process [14].

3D digitizing, and other sources of 3D data. SLS has certain advantages such as the production of large and complex parts. Small batches can also be produced in a single manufacturing process without the need for supports because undercuts and overhangs are supported by a solid powder bed. Parts can also be finished to any degree, are watertight and are autoclave sterilizable. The process provides good parts accuracy compared to the FDM process.

The process can be used for fabricating patient-specific anatomical models that can be used in dentistry, orthopedics, and neurological surgery. It can also be used for implant fabrication.

A stereolithography apparatus (SLA) is a vat photo polymerization–based AM process. This method makes use of photochemical/photosensitive materials through which an ultraviolet laser passes to initiate the polymerization process. Figure 9.5 is a schematic diagram of the SLA process. Due to the restricted availability of SLA processable materials, the application is also limited. The process can be used to fabricate complex anatomical models and replicas and for presurgical planning with high resolution and accuracy.

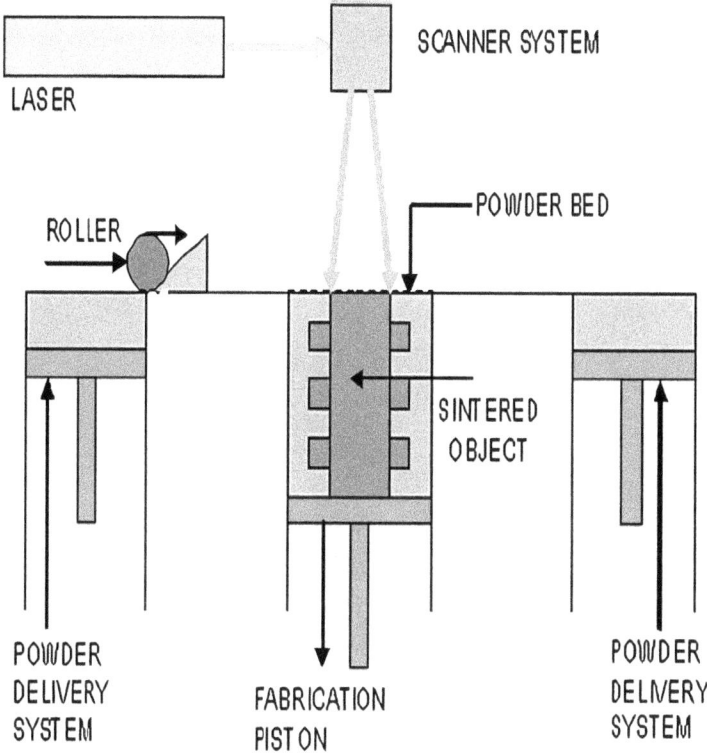

FIGURE 9.4 Schematic of the SLS process [15].

Selective laser melting (SLM) is very similar to SLS with the exception of processing metals using high-energy fiber lasers. The process completely melts the metal powder and is capable of producing parts with high accuracy and a good surface finish. The advantage of processing metals opens up a wider range of biomedical applications such as prostheses, knee and hip implants, dental implants, and surgical guides. Figures 9.6 and 9.7 illustrate the concept of the SLM process and various applications for metal AM in the biomedical sector.

9.3 SOLID MODELING AND BIOMATERIALS FOR AM PROCESSES

For solid modeling of the anatomical details of parts, a CT or MRI scan is used. The details can easily be decoded by computers and converted into a point cloud. The point cloud is further processed for the STL model. The segmentation of the medical data is a crucial task that requires expert knowledge to segment the intended part of the CT scan file. CT scan files are in DICOM format. Each pixel of the DICOM image has a certain value that corresponds to the x-ray attenuation coefficient of the body part. Parts with high density, i.e., bone, have higher x-ray attenuation, hence they reflect the color on the bright side of grayscale; on the other hand, body parts

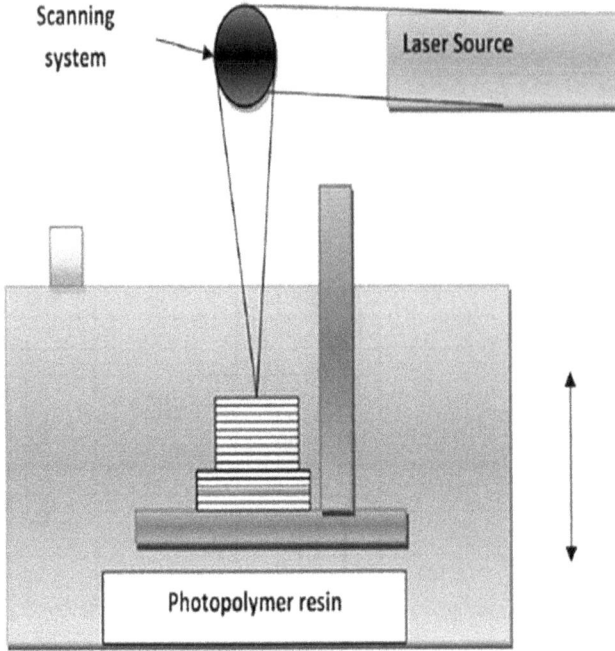

FIGURE 9.5 Schematic of the SLA process [16].

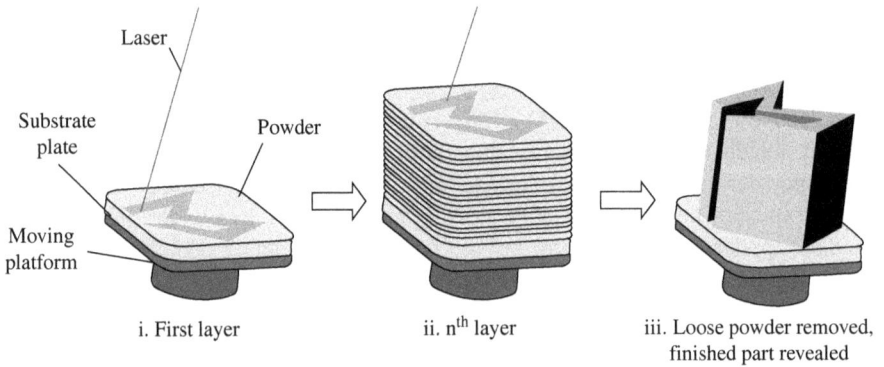

i. First layer   ii. n^{th} layer   iii. Loose powder removed, finished part revealed

FIGURE 9.6 Schematic of the SLM concept [17].

that have less density attenuate the x-rays and reflect the color on the darker side on the CT scan. Thus, on the bases of the gray value of the pixel, it can be segmented out for different body parts. Based on the grayscale values, different parts can be segmented out and a mask created. This mask is the segmented part of the 3D array of the CT scan or MRI scan images. The generated mask is interpolated to convert the point cloud into a voxel model. The voxel model is used to render the solid geometry on the computer screen. Furthermore, this point cloud data is converted into an STL

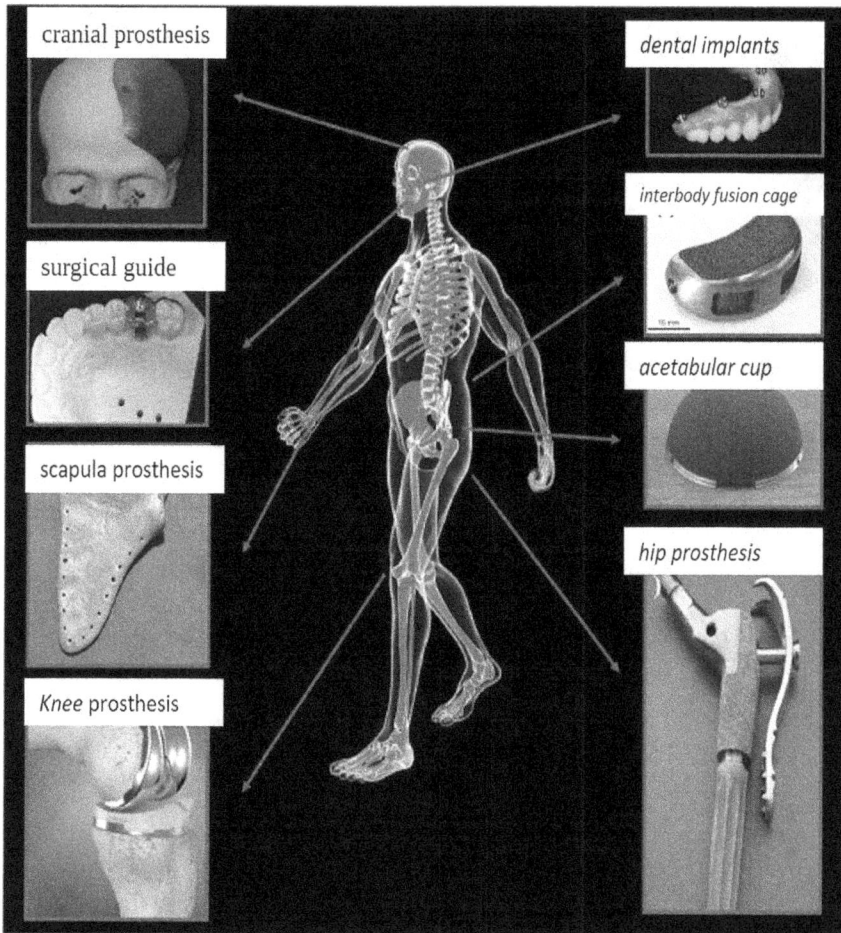

FIGURE 9.7 Some biomedical applications of metal AM [18].

file. The STL file is then used for 3D printing. Figure 9.8 illustrates the segmentation of CT scan images and their 3D model.

Computational software makes it easy for clinicians to develop a solid model of implants and surgical guides for 3D printing. Advanced 3D printing techniques such as selective laser melting, stereolithography, and poly jet make it convenient and reliable to use in medicine.

FDM-based 3D printing is used in guided endodontics to create a root canal access cavity for calcified canals [8]. The results of guided endodontics are promising and encourage the adoption of the latest 3D printing and designing technology in medicine. Additive manufactured plates are used in the reconstruction of the mandible and maxilla [9]. Moreover, 3D-printed guides are used to fabricate the cutting guides for resectioning purposes.

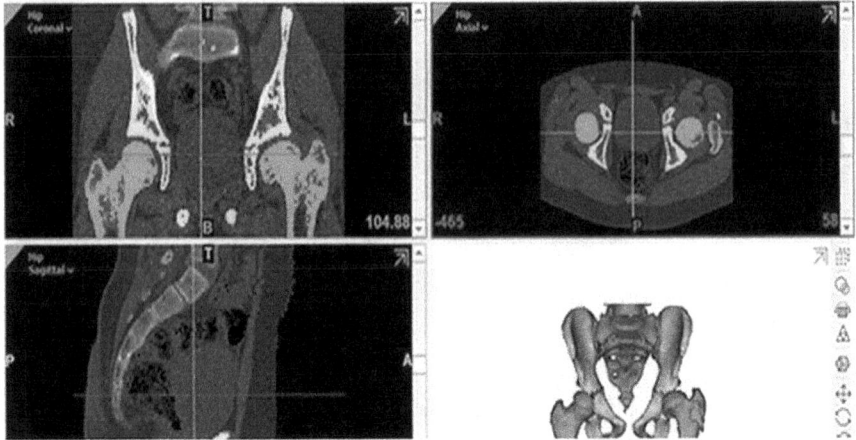

FIGURE 9.8 Segmentation of CT scan images and voxel model representation.

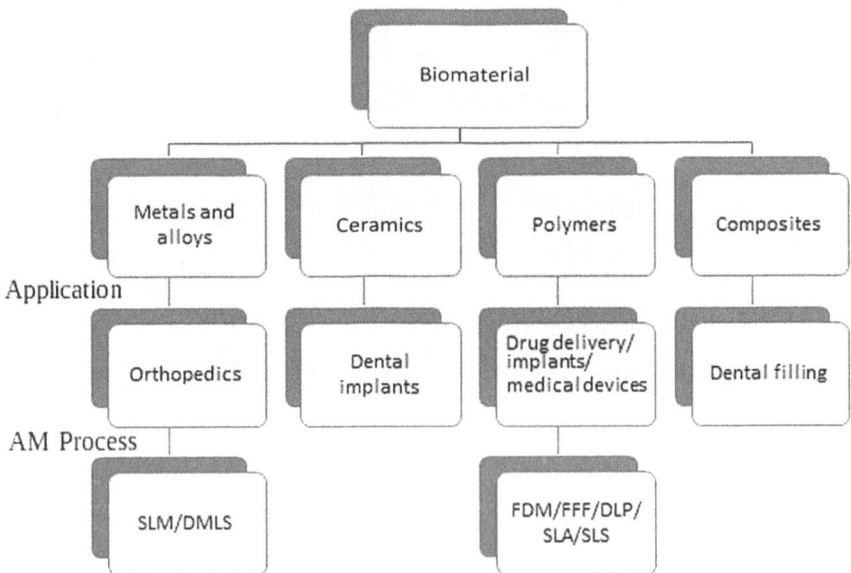

FIGURE 9.9 Classification of biomaterials used for AM [19].

Materials play an important role in biomedical applications. The selection of materials depends upon the final part requirement. To print the biomaterial via AM, firstly it should be printable and most importantly biocompatible. Metals, cremains, and polymers are used for various biomedical applications. Figure 9.9 shows the classification of biomaterials and various AM techniques that can process them.

9.4 AM APPLICATIONS AND OPTIMIZATION OF PARAMETERS

AM is utilized for various applications in orthopedics. The fabrication of anatomical models used to plan surgical procedures is one such application. Using AM physical models, surgeons can gain a visual and tactile understanding of the disease and anatomy specific to a patient. Traditionally, this is done by 2D x-ray images or 3D CT scan images using processing software. Surgical guides can be fabricated by AM for orthopedic applications. These products are customized to the patient's requirements. Moreover, they cost less because the cost associated with tooling is eliminated. Patient-specific instruments can be fabricated as per the anatomy of the patient, which aids surgery. Bone tissue engineering is another application where AM is utilized for orthopedics. AM helps in the design and fabrication of bioscaffolds as per the clinical requirements. A weight reduction in implants can be achieved using the AM process because it allows the design of complex parts which can help to reduce the weight of implants [10,11].

One area of medicine where 3D printing has had a significant impact is dentistry. Due to the process's capacity to produce prosthetic teeth and dental implants quickly, sometimes even inside a dentist's office, creating them has become simpler. Obtaining the picture of the organ is the first step in the process of 3D manufacturing any implant. The implant is created using an image of the organ as a basis, and it is then layer-by-layer constructed [12].

AM is widely used for the fabrication of maxillo-facial implants using materials such as titanium and calcium phosphate. Mechanical performance similar to teeth can be achieved using the AM process. Corrosive-resistant dentures and crowns can also be easily fabricated using AM. Materials such as cobalt chrome, plastics, alumina, and zirconia are used in the fabrication of dentures and crowns. AM is also utilized for the fabrication of anatomical and training models in the field of dentistry. Furthermore, scaffolds are fabricated to achieve the same color and mechanical properties as teeth.

The process parameters of AM need to be optimized for better performance of biomedical implants and other fabricated components. Each AM process has different process parameters that can significantly affect the fabricated parts. Lased-based AM processes such as SLA and SLM have laser power, scan speed, build orientation, hatch spacing, bed temperature, and layer thickness as major contributing parameters. These can affect the surface refinish and dimensional accuracy of implants. In an extrusion-based process such as FDM, parameters such as layer thickness, infill density, infill pattern, orientation, extruder and bed temperature, and even geometries are optimized.

9.5 CONCLUSION

AM has emerged as a major contender in the field of biomedicine for the fabrication of various implants, devices, and scaffolds. This is due to the fact that any complex features that are majorly associated with the biomedical sector can be easily fabricated with AM. Moreover, fabrication in most cases takes place in a single setup and a short time period. The recent advances in biomaterials that can be processed by AM have further enhanced its use for biomedical applications. Various materials such as polymers, ceramics, and metals have been used in the manufacture of several biomedical components and devices using AM with the required properties. It can be concluded that AM possesses great potential for use in various biomedical applications.

REFERENCES

1. Singh S., S. Ramakrishna. Biomedical applications of additive manufacturing: Present and future. *Current Opinion in Biomedical Engineering* 2017;2, 105–115.
2. Francis V. et al. (2020). Influence of 3D printing technology on biomedical applications: A study on surgical planning, procedures, and training. In Singh, S., Prakash, C., Ramakrishna, S., Krolczyk, G. (eds) *Advances in Materials Processing. Lecture Notes in Mechanical Engineering*. Springer, Singapore, 269–278.
3. Ganguli A., G.J. Pagan-Diaz, L. Grant, C. Cvetkovic, M. Bramlet, J. Vozenilek, T. Kesavadas, R. Bashir. 3D printing for preoperative planning and surgical training: A review. *Biomedical Microdevices* 2018;20(3), 65.
4. Ukey P., P.K. Jain, R.V. Uddanwadiker. (2016). Fabrication of artificial temporal bone from CT Scan data using FDM technique for dissection training. Unpublished Dissertation. PDPM Indian Institute of Information Technology Design and Manufacturing, Jabalpur, Madhya Pradesh, India.
5. Taufik M., P.K. Jain. A study of build edge profile for prediction of surface roughness in fused deposition modeling. *Journal of Manufacturing Science and Engineering* 2016;138(6), 061002.
6. ASTM International. ISO/ASTM52900—15 standard terminology for additive manufacturing—General principles terminology. ASTM International, West Conshohocken, PA, 2015.
7. Salmi M. Additive manufacturing processes in medical applications. *Materials* 2021;14(1), 191.
8. Nayak A., P.K. Jain, P.K. Kankar, N. Jain. Computer-aided design–based guided endodontic: A novel approach for root canal access cavity preparation. *Proceedings of the Institution of Mechanical Engineers, Part H: Journal of Engineering in Medicine* 2018 Aug;232(8), 787–795.
9. Su R.Y. A prospective clinical trial on 3D-printed patient-specific titanium plates in head and neck reconstruction: Clinical outcomes, efficiency and accuracy results. *Journal of Oral and Maxillofacial Surgery* 2020 Oct 1;78(10), e95–96.
10. Javaid M., A. Haleem. Additive manufacturing applications in orthopaedics: A review. *Journal of Clinical Orthopaedics and Trauma* 2018 Jul–Sep;9(3), 202–206.
11. Ejnisman L., B. Gobbato, A.F. de França Camargo, E. Zancul. Three-dimensional printing in orthopedics: From the basics to surgical applications. *Current Reviews in Musculoskeletal Medicine* 2021 Feb;14(1), 1–8.
12. Bhargav A., V. Sanjairaj, V. Rosa, Lu Wen Feng, Jerry Fuh YH. Applications of additive manufacturing in dentistry: A review. *Journal of Biomedical Materials Research, Part B – Applied Biomaterials* Jul 2018;106(5), 2058–2064.
13. Gross B.C., J.L. Erkal, S.Y. Lockwood, C. Chen, D.M. Spence. Evaluation of 3D printing and its potential impact on biotechnology and the chemical sciences. *Analytical Chemistry* 2014;86(7), 3240–3253. https://doi.org/10.1021/ac403397r.

14. Do A.V., B. Khorsand, S.M. Geary, A.K. Salem. 3D printing of scaffolds for tissue regeneration applications. *Adv Health Care Mater* 2015;5, 44–59. https://doi.org/10.1002/adhm.201500168.
15. Stansbury J.W. Idacavage M. J. 3d printing with polymers: Challenges among expanding options and opportunities. *Dental Materials* 2016;32, 54.e64. https://doi.org/10.1016/j.dental.2015.09.018.
16. Awasthi A., K. K. Saxena, R. K. Dwivedi. An investigation on classification and characterization of bio materials and additive manufacturing techniques for bioimplants. *Materials Today: Proceedings* 2021;44(Part 1), 2061–2068.
17. Yap C.Y., C.K. Chua, Z.L. Dong, Z.H. Liu, D.Q. Zhang, L.E. Loh, S.L. Sing. Review of selective laser melting: Materials and applications. *Applied Physics Reviews* 2015;2(4), 041101. https://doi.org/10.1063/1.4935926.
18. Ni J., H. Ling, S. Zhang, Z. Wang, Z. Peng, C. Benyshek, R. Zan, A.K. Miri, Z. Li, X. Zhang, J. Lee, K.-J. Lee, H.-J. Kim, P. Tebon, T. Hoffman, M.R. Dokmeci, N. Ashammakhi, X. Li, A. Khademhosseini. Three-dimensional printing of metals for biomedical applications. *Materials Today Bio* 2019;3, 100024. ISSN 2590-0064.
19. Kumar R., M. Kumar, J. S. Chohan. The role of additive manufacturing for biomedical applications: A critical review. *Journal of Manufacturing Processes* 2021;64, 828–850.

10 Scope of Micromanufacturing in Medical Implants

Vinod Yadav and Farheen Khan

10.1 INTRODUCTION

In the field of micro-level miniaturization, where medical implants can benefit both economically and technically, micromanufacturing is quickly becoming a significant manufacturing technique. Micromanufacturing of medical implants has a fast-growing market having a broad spectrum of biomedical applications. The report published by the World Health Organization (WHO) in 2020 stated that there is an urgent need to enhance the methods of manufacturing technologies of medical implants at the micro level. Due to its compactness, minimal material requirement, low power consumption, great sensitivity, and several other advantages over large items, medical implants are in high demand.

A component can be manufactured using micromanufacturing techniques if at least one orthogonal view can fit inside a square of 1 mm in size [1]. Jain et al. [2] mentioned that manufacturing sub-millimeter-size features on small or large components is also known as a micromanufacturing process. For example, the diameter of a dental implant screw can be more than 1 mm, but one tooth may be much smaller. Some of the most commonly implanted medical devices include artificial joints, breast implants, cochlear implants, intraocular lenses, pacemakers, and other cardiac implants.

Micromachining, microforming, additive manufacturing (AM), mass containing, micro-joining, and nanofinishing are some of the subcategories of micromanufacturing processes. Small size material chips are removed during micromachining in order to create the desired form, size, or feature. Micromachining processes are further subcategorized into micro-turning, micro-drilling, and micro-milling as per the specimen geometry and the operations required to convert the raw material into product form. Microforming processes are categorized into bulk microforming and sheet microforming. Examples of bulk microforming are micro-rolling, micro-extrusion, and micro-forging, and examples of sheet microforming processes are micro-deep-drawing and micro-bending. Laser metal deposition, selective laser melting (SLM), 3D printing, and cladding are examples of additive manufacturing processes. The microcasting process is an example of mass containing. Micro-welding and

DOI: 10.1201/9781003375098-10

micro-bonding are two examples of micro-joining. In order to achieve a high level of accuracy and surface finish, nanofinishing techniques are used. Figure 10.1 shows a typical comparison of the surface topology obtained using conventional manufacturing and the additive 3D printing technique of a knee joint, hip socket, and stent implants. It can be seen from Figure 10.1 that 3D printing technology provides accurate dimensions of the desired product; however, further operations may be needed to improve the surface roughness.

The manufacturing process becomes more complicated at micro-level manufacturing due to the presence of size effect. An underformed chip thickness reduced to levels below the cutting-edge radius or the grain size of the workpiece material, a phenomenon that occurs in micromanufacturing, is called the size effect [1,2]. Balasubramanian and Suri [3] reported that when the uncut chip thickness is lowered from a few micrometers to submicrometers on a brass workpiece with a polycrystalline diamond (PCD) tool of 100 μm tool nose radius (TNR) and a 1.25 cutting-edge radius, a dramatic rise in the specific cutting energy occurs when the uncut chip thickness falls below the cutting-edge radius.

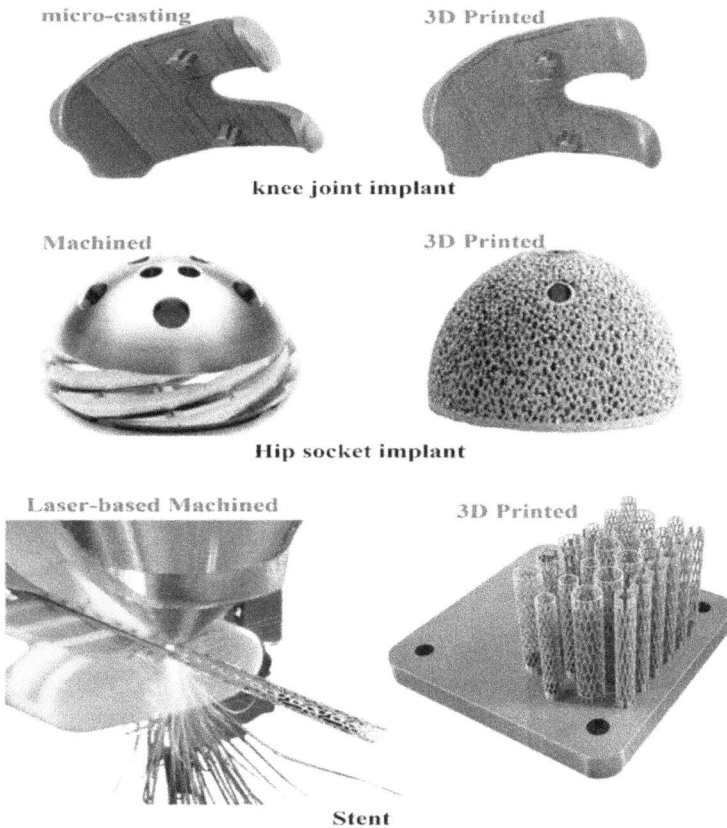

FIGURE 10.1 Typical manufactured medical implants using different methods.

The chapter is organized as follows. Section 10.1 introduces the importance of micromanufacturing in medical implants and their applications. The different types of medical implants commonly used in the health sector are discussed in Section 10.2. Section 10.3 presents the materials used to manufacture medical implants. Section 10.4 discusses the different micromanufacturing methods employed in the manufacturing of medical implants. Section 10.5 concludes the chapter with directions for future research.

10.2   MEDICAL IMPLANTS

Medical implants are temporary or permanent man-made devices that are used in the body for diagnostic, monitoring, or therapeutic purposes. Implants also deliver medication and can provide support to organs and tissues. Figure 10.2 shows the various types of medical implants placed inside the human body or pasted onto the surface of the human body. These are broadly classified into seven types of manufactured medical implants: breast, cerebral spinal fluid shunt (CSFS) systems, essure permanent birth control, hernia surgical mesh, and hearing, hip, and phakic implants.

Figure 10.3 shows the different types of medical implants that are pasted onto the surface of the human body and inserted inside the human body. Breast implants enlarge and change the shape of the breast. Breast implants are placed behind the breast tissue or beside the chest. They are made from either saline water or silicon material [4]. There are three types of breast implants: saline-filled breast, silicon-filled breast, and breast tissue expanders. Saline implants are filled with sterile salt water in the empty space. Deflation occurs when the breast is underfilled with saline [5]. Silicone gel fillings are used in silicone implants. Saline-filled breast implants are better than silicone gel–filled breast implants [5]. Inflatable implants, known as breast tissue expanders, stretch the skin and muscle to make room for permanent implants [6].

A shunt is a flexible tube that is inserted into the ventricle, a region of the brain where cerebrospinal fluid (CSF) amasses. The shunt's objective is to remove the additional fluid that is raising the pressure inside the brain. There are two types of

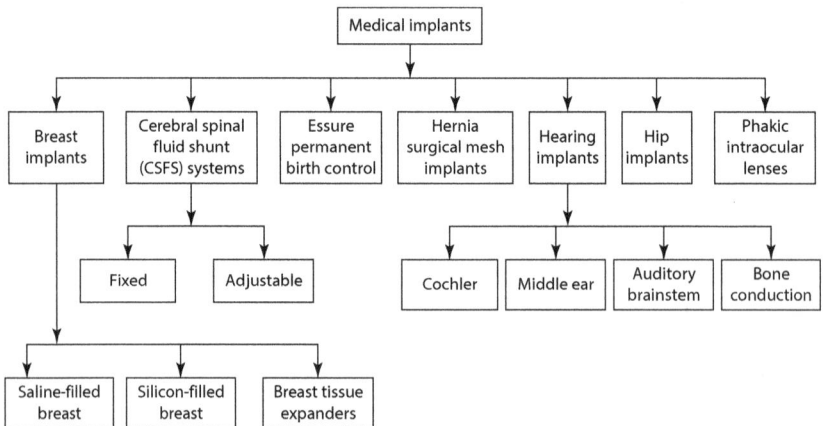

FIGURE 10.2   Types of medical implants.

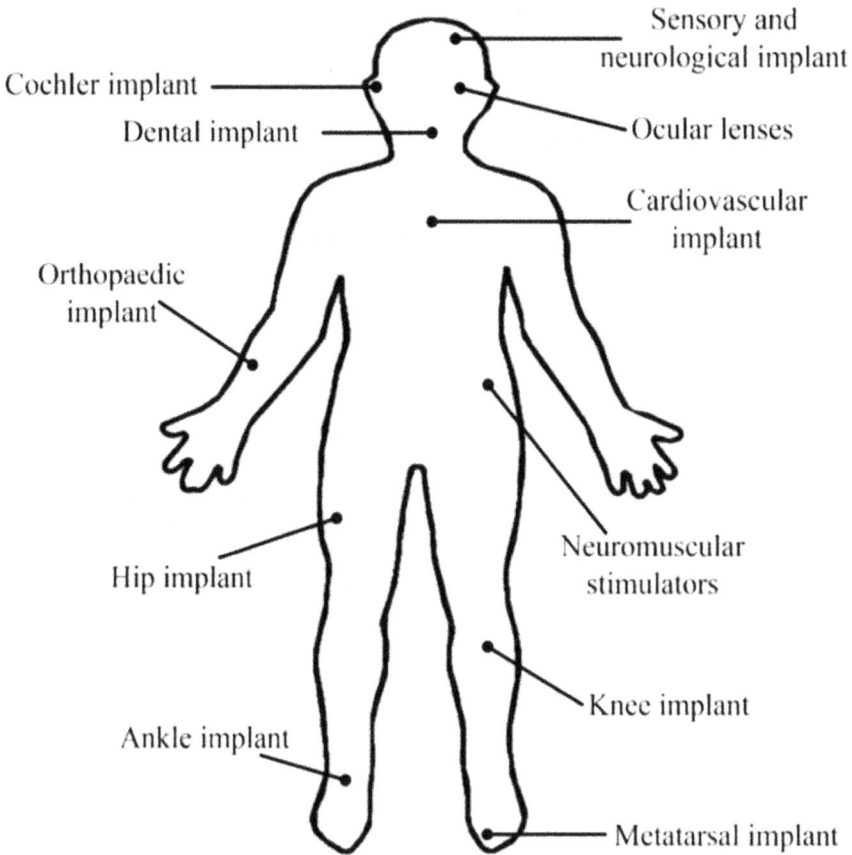

FIGURE 10.3 Schematic representation of various implants placed inside or on the surface of the human body.

CSF shunt valves: fixed and adjustable. If the CSF pressure rises over a certain set threshold, fixed shunt valves enable CSF fluid to drain. The amount of fluid that passes through adjustable shunt valves is increased or decreased. There are two types of adjustable CSF shunt valves: magnetic externally adjustable and non-magnetic externally adjustable [7]. Essure is recommended for women who want female sterilization (permanent birth control) by bilateral fallopian tube closure. Each of the patient's fallopian tubes receives a soft, flexible insert, and over the course of three months, a barrier develops around the inserts to stop conception.

The bulk of the surgical mesh devices that are currently used are made of synthetic materials or animal tissue [4]. Pain, infection, hernia recurrence, adhesion, and intestinal obstruction are the most frequent side effects following mesh hernia treatment. Mesh migration and mesh shrinkage (contraction) are two additional possible side effects that may appear after mesh hernia surgery. A cochlear implant, a tiny, highly advanced electrical device, allows someone who is totally deaf or has very limited hearing to hear sound. The exterior component of the implant is located behind the ear, while a second section is surgically inserted beneath the skin. An implant for bone conduction transmits sound directly to the inner ear, successfully avoiding the outer and middle ear. The inner ear is mechanically stimulated via a middle ear implant that is inserted into the middle ear. It is intended for those who are medically unable to wear hearing aids. For patients with severe hearing loss who cannot utilize a cochlear implant, auditory brainstem implants are available. Similar to the cochlear implant, the auditory brainstem implant employs modern engineering.

Hip implants are devices that are used to increase mobility and reduce discomfort from hip disorders or accidents, such as arthritis [7]. Every hip implant has advantages and disadvantages. Size, form, material, and dimensions are just a few of the unique device design features that each hip implant system possesses. In order to lessen the need for glasses or contact lenses, phakic intraocular lenses, also known as phakic lenses, are lenses composed of silicone or plastic that are permanently implanted into the eye. Phakic describes the process of implanting a lens into the eye without removing the normal lens of the eye. Through the cut, the phakic lens is introduced and positioned immediately in front of or just behind the iris [8].

10.3   MATERIALS USED FOR MANUFACTURING MEDICAL IMPLANTS

A number of materials are used in the manufacturing of medical implants such as metallic alloys, polymer, composite, and ceramics [4,8].

10.3.1   METAL AND ITS ALLOYS

Davis et al. [4] presented a review on biomaterials for metallic implants and their different manufacturing processes. They mentioned that common metals and their alloys are used in the manufacturing of medical implants: stainless steel (SS), titanium (Ti)-based alloys, cobalt-chromium (Co-Cr)-based alloys, nickel-titanium (Ni-Ti)-based shape memory alloys (SMA), magnesium (Mg)-based alloy, and bulk metallic glass (BMG). For load bearing and orthopedic applications, biomedical implants must also have excellent yield, fatigue, tensile, compressive, and shear strength in addition to good biocompatibility [9].

10.3.1.1   Stainless Steel

Metallic surgical implants accelerate bone consolidation following fractures. Compression plates held against the bone by bolts and nuts comprise an implant group. Traditional methods (without implants) could cause cartilage and articulation atrophy after an excessively long consolidation period. The metals and alloys used in osteosynthesis implants have been studied in recent years to improve their biocompatibility. Medical devices made from them are economical and easy to manufacture. The stiffness of stainless steel implants is ten times higher than human bone [10]. Moreover, conductive oxides can cause inflammation. Despite this, nickel has superb mechanical properties. Compared to other stainless steels for implant manufacturing, it is stronger when annealed [10].

10.3.1.2 Ti-Based Alloys

Titanium is bio-inert, causing little or no damage to surrounding tissue. According to studies on dental implants, titanium oxide (TiO_2) has the most reported properties among titanium forms. Due to titanium's excellent capacity to react with air to form hydroxyl and hydroxide groups, greatly increasing its ability to prevent corrosion, TiO_2 was created [11]. Due to its advantageous characteristics, which include excellent biocompatibility, mechanical attributes, wear and corrosion resistance, a high strength-to-weight ratio, reduced stiffness, and low elastic modulus, commercially pure titanium and Ti-6Al-4V alloy, a biomedical-grade titanium and its alloys have consistently been among the most dependable choice of users despite their high cost [12].

10.3.1.3 Co-Cr-Based Alloys

Due to their notable resistance to corrosion and wear properties, cobalt-based alloys have gained in popularity for orthopedic implantation. Co-based alloys can be classified into two types, each containing approximately 28% chromium and 5% molybdenum [13]. Significant advancements have been made in the development of cobalt-based metallic implants to mitigate total volumetric wear. Co-Cr alloy implants, for instance, have practical applications in the fields of dentistry and orthopedics, similar to Ti alloys. However, Co-Cr alloy implants with elevated nickel content may lead to allergic reactions and lower biocompatibility with the host tissue [14,15].

10.3.1.4 Ni-Ti Shape Memory Alloy

In recent years, an alloy known as nickel-titanium shape memory alloy has been created by combining alloys of nickel and titanium in roughly equiatomic proportions. Ni-Ti is a member of the family of shape memory alloys, which can be bent at low temperatures and then recovered at high temperatures. The Ni-Ti shape memory alloy, which was formerly known as nitinol, is non-cytotoxic, lightweight, and has great strength and corrosion resistance. Nitinol implants help with a quicker recovery of bone than SS, Ti, and Co-Cr implants due to their capacity to reduce stiffness in the precise proportion of a bone. This has led to superior load-bearing throughout the healing process [16–18]. A high nickel intake might result in cancer and allergic responses [19].

10.3.1.5 Mg-Based Alloy

Due to their considerable potential for use as temporary orthopedic implants, alloys based on magnesium have emerged as a significant category of materials. In orthopedics, these alloys are an effective substitute for implants made of non-degradable metal [20]. Examples of non-degradable and biodegradable implants include dental implants, heart pacemakers, and cardioverter defibrillators, spine screws, rods, and discs. The surface of a tailored biomedical implant interacts with the blood, bodily fluids, tissues, proteins, and even cells at the microscopic level after it is placed in the human body. The patient is relieved of the anxiety associated with secondary surgery if the implant can be removed [21,22]. Additionally, the combination of Mg with Ag, Ca, Cu, and Sr speeds up the modeling of cells and bone growth [23]. Figure 10.4 depicts the degradation behavior of Mg-based alloy implants inserted in the bone of a human body.

10.3.1.6 Bulk Metallic Glass

Over the last three decades, a relatively new family of metallic materials known as bulk metallic glasses has been produced [25]. Because of its advantageous hardness, tensile strength, elasticity, minimal internal friction, and adequate resistance to corrosion and wear, bulk metallic glass has also been regarded as a viable material for biomedical implants. However, the lack of crystallinity makes BMG difficult to produce. Su et al. [26] suggested three of the most important forming criteria in the manufacture of a BMG implant: (i) requires an alloy with three or more different elements; (ii) indicates an atomic size discrepancy of at least 12% between the main parts; and (iii) mixing between the main components at a negative temperature.

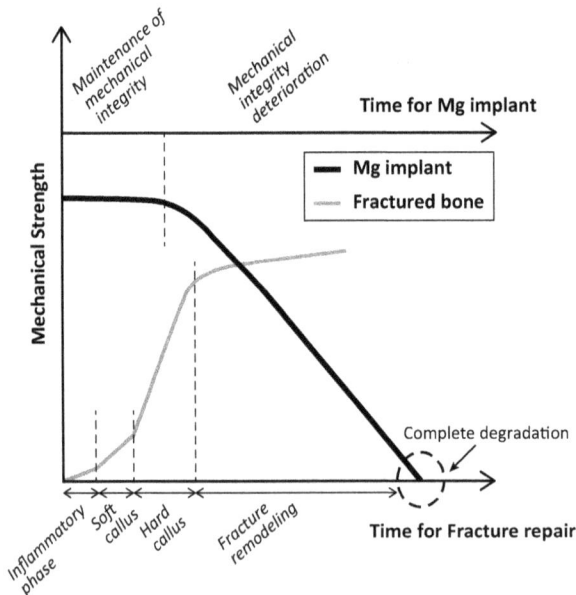

FIGURE 10.4 Degradation behavior of Mg-based temporary implants in the bone fracture healing process under ideal conditions. With permission from Zhao et al. (2017) copyright (2017) Elsevier.

10.3.2 POLYMER

Small, cylindrical rods having a length of around 1–1.5 cm and a diameter of 1–2 mm are known as polymeric implants. They often include the same polymeric components. For the administration of implants, large needles (e.g., with a gauge of 16) or surgical incisions are necessary [27]. Due to their strength, inertness, and biocompatibility, polymer biomaterials have been used in clinical practice to relieve the load of damaged or infected bone and enhance the patient's quality of life. Polymer biomaterials are used in a variety of tissues, such as the cardiovascular [28], neural [29], musculoskeletal [30], and dermal [31] tissues. Ultra-high molecular weight polyethylene, polyethylene, high density polyethylene, and polyacetal are a few examples of polymer implants.

10.3.3 COMPOSITE

Composite material implants are utilized for central bone replacement and joint and bone fusions. The main goal is to create a composite implant consisting of fibers that can stimulate more bone formation and have less micromotion than the titanium alloy (Ti6Al4V) implants that are now in use [32]. Orthopedic implant materials made of carbon fiber/PEEK polymer (C/PEEK) composite materials are currently being developed. Wear is an increasing problem that affects orthopedic implants; particle debris produced by biomaterial wear may be a cause of osteolysis and implant loosening. In order to characterize the wear of C/PEEK composite materials in contrast to contemporary orthopedic implant materials, numerical and experimental experiments were undertaken [7].

10.3.4 CERAMICS

Researchers and physicians requirements were not addressed by polymer-based or metal-based implants, so they turned to ceramic as a substitute. Because of their great hardness, high wear resistance, and excellent biocompatibility, ceramic materials are often suitable for bone replacement bearings. Compared to metals and polymer implants, ceramics produce a very small quantity of wear debris [33]. In order to meet the expectations for an aesthetically pleasing metal-free treatment for edentulous jaws, all-ceramic dental implants have been discovered as a viable substitute for the conventional titanium-based implant systems used in dentistry. Zirconia implants have become the standard among all-ceramic implants used in dentistry. Few clinical trials have evaluated zirconia's long-term success, despite the fact that its short-term implant success has mirrored that of titanium. Zirconia dental implants have been known to fail due to poor operator technique, manufacturing flaws, and unfavorable loading. As a result, manufacturers must take the best possible quality control measures, and an experienced operator must plan an effective course of treatment to maximize the benefits of zirconia as an implant biomaterial [34]. The application of different implant materials used to manufacture implants along with their mechanical properties is shown in Table 10.1.

TABLE 10.1
Implant Materials, Mechanical Properties, and Their Applications

Implants material		Implants manufactured	Yield stress (MPa)	Hardness (HV)	References
Metal and its alloys	Stainless steel	Orthopedic, dental, cardiovascular	220–260	130–160	Yang and Ren (2010)
	Ti-based alloy	Dental and orthopedic	951	1800	Rack and Qazi (2006)
	Co-based alloy	Hip and knee joint	450–520	240	Singh et al. (2016)
	Ni-Ti shape memory alloy	Cardio stents	70–140	346	
	Mg-based alloy	Bone screws, plates, stents	200	65	Sankaranarayanan et al. (2015)
	Bulk metallic glass	Stents, screws, pins, plates	1000–4000	—	Meagher et al. (2016)
Polymer		Hemodialysis membranes	—	—	Davis et al. (2016)
Composite		Bone fracture repair, hip joint, knee joint	—	—	
Ceramic		Joint repair	—	—	

10.4 MICROMANUFACTURING METHODS OF MEDICAL IMPLANTS

Commonly, medical implants are manufactured from the materials listed in Section 10.3 due to their biocompatibility, high strength-to-weight ratio, and great resistance to corrosion and wear [35]. Among those materials, Ti-based alloys and Co-Cr-based alloys face a series of difficulties related to their low machinability during the manufacture of medical implants. According to a literature assessment, titanium-based alloys and stainless steel make up the majority of commercially available implants [36]. One of the least expensive implants for biomedical use is stainless steel. Such implant materials must be machined to the desired form and size using either traditional micromanufacturing techniques alone, non-traditional micromanufacturing methods in combination, or both. Machining, forming, casting, and joining are examples of traditional manufacturing processes. By contrast, electrostatic discharge machining (EDM), wire electric discharge machining (WEDM), abrasive water jet machining (AWJM), ultrasonic machining (USM), and cryogenic machining are emerging technologies. Figure 10.5 depicts a fishbone diagram showing the performance of medical implants through micromanufacturing techniques. Various micromanufacturing techniques are employed for different medical implants in order to achieve the desired mechanical properties and surface roughness characteristics. Process parameters such as biocompatibility, eligibility (age of patient), durability, and friction between the tissue and implanted

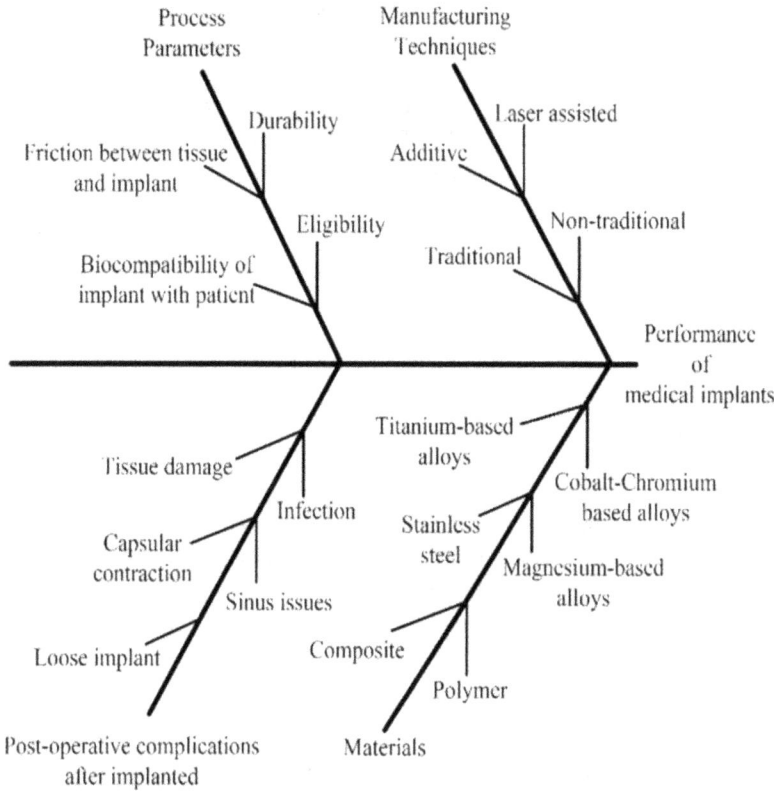

FIGURE 10.5 Fishbone diagram of the performance of medical implants.

devices play a key role in selecting the implant materials prior to manufacturing. Post-operative complications should be recorded to estimate the life span of the medical implants.

10.4.1 TRADITIONAL MICROMANUFACTURING PROCESSES

In order to enhance the machinability of medical implants made of Ti-based alloys, Co-based alloys, and stainless steel, traditional micromanufacturing processes are employed. For example, Abellan-Nebot et al. [35] envisaged the manufacturing process plan for the metallic components of a prosthesis. In traditional manufacturing processes, there are three steps in the manufacture of a prosthesis implant. In the first step, casting and sintering are employed to create the structure of the implant. Shaping is done through forging. Subsequently, for the desired shape, size, and features, machining processes such as milling, turning, and grinding are employed. Lastly, a finishing process is used to achieve the desired surface roughness of the implant. In terms of surface quality and productivity, the surface roughness is one of the most important factors in the field of medical implants.

Danish et al. [37] experimentally investigated the effect of both cryogenic and dry micro-turning of medical implants on surface roughness and temperature generation. The workpiece material was considered an Mg-based alloy. The cryogenic micro-turning experiments were performed by applying liquid nitrogen at the cutting tool and workpiece interface. They found that the surface roughness and microstructure were greatly affected by the temperature generated during the cutting of Mg-based alloy implants. Thus, the temperature affects the corrosion properties of the medical implants of an Mg alloy. They found that the temperature of the chip and the cutting tool and workpiece interface is reduced during cryogenic machining as compared to that during dry machining.

Zhao et al. [38] analyzed the chip formation behavior of a Ti-based alloy workpiece material in cryogenic cutting and dry cutting at different input process parameters. Dry machining was performed at room temperature whereas cryogenic machining was in the range of –40µC to 196µC. They found that cryogenic cutting is superior to that of dry cutting due to the low magnitude of the temperature generated. As a consequence, the chip height ratio and serrated pitch were greatly increased, thereby enhancing the machining performance of Ti-based alloys. The degree of chip deformation is characterized by the chip height ratio measurement. Figure 10.6 is a schematic representation of the measurement of the chip height ratio and serrated pitch. The chip height (G_s) can be calculated as

$$G_s = \frac{H - h}{H} \tag{10.1}$$

where H is the height from the chip's base to its highest point and h is the distance from the chip's base to its deepest valley. The serrated pitch can be measured by measuring the distance between two adjacent highest peaks. It was found that the

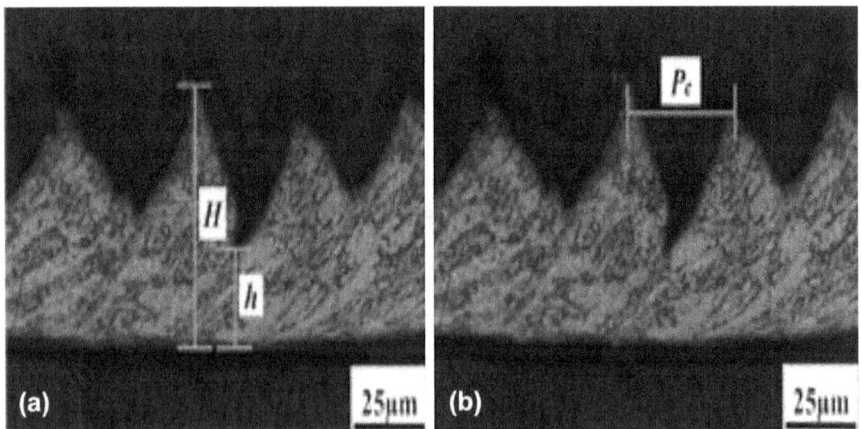

FIGURE 10.6 Schematic representation of chip height and serrated pitch measurement of Ti-based alloys. With permission from Zhao et al. (2018) copyright (2018) Elsevier.

chip height ratio and serrated pitch were increased by 45.2% and 58.3% during cryogenic machining as compared that of dry machining.

10.4.2 Non-Traditional Micromanufacturing Processes

Non-traditional manufacturing processes are widely used to machine hard and difficult-to-cut materials with complex shapes. To increase the surface integrity and productivity of medical implants, these processes are most effective for improving the surface roughness and material removal rate of medical implants. Using the WEDM process, Raju et al. [39] machined orthopedic implants made of stainless steel. Considering the surface roughness of the machined specimen as an output performance parameter, the effects of pulse on time, peak current, servo voltage, and wire tension were studied. They found that the process parameter pulse on time is more pronounced on the surface roughness achieved through WEDM of medical implants. Trimble et al. [40] examined the flow stresses of a biomedical-grade Co-27Cr-5Mo alloy as a function of strain, strain rate, and temperature. They carried out split-Hopkinson pressure bar (SHPB) experiments over a wide range of domain temperatures (298–873 K) and strain rates (600–1400 s^{-1}), and then they fitted the Johnson–Cook and Arrhenius-type constitutive equations to obtain the material constants by minimizing the error between the experimental and the predicted data. Co-27Cr-5Mo alloy cylindrical specimens were machine-prepared using WEDM from samples of knee implants made from tibia trays using investment casting. They found that the Arrhenius-type constitutive model in the deformation behavior more accurately of biomedical grade Co-27Cr-5Mo alloys in traditional machining process.

Hard machining of brittle metallic medical implants such as ceramics is a challenging task. Overcoming such challenges, Putz et al. [41] employed two different techniques viz., injection-type AWJM and suspension-type AWJM processes, to make a cut or hole in the orthopedic implants of titanium and stainless alloys. In these processes, the two different jet generation methods are employed. Figure 10.7 is a schematic representation of two different abrasive water jet (AWJ) processes at the macro and microscale. Putz et al. [41] reported that the AWJ process showed significant potential and a promising technique to produce micro channels. Figure 10.7 shows the schematic representation of two different abrasive water jet (AWJ) process at macro and micro scale. At macro level surface finishing has been with AWJ process principle shown in Figure 10.7a whereas Figure 10.7b shows the surface finishing operations to micro range.

Estimation of the life span of a medical implant is between 15 and 20 years. The rough surface attracts blood-clotting proteins [42]. To enhance the life span of medical implants (e.g., hip joint implant, knee joint implant, and shoulder joint implant), the surface finish needs to be produced at the nanoscale level. For this particular phenomenon, the abrasive flow finishing (AFF) process plays a key role in producing a nanoscale surface finish without altering the metallurgical properties of medical implants. Ravisankar et al. [42] reported surface finish improvements in a hip joint and a knee joint implant placed in a human knee before and after the AFF procedure.

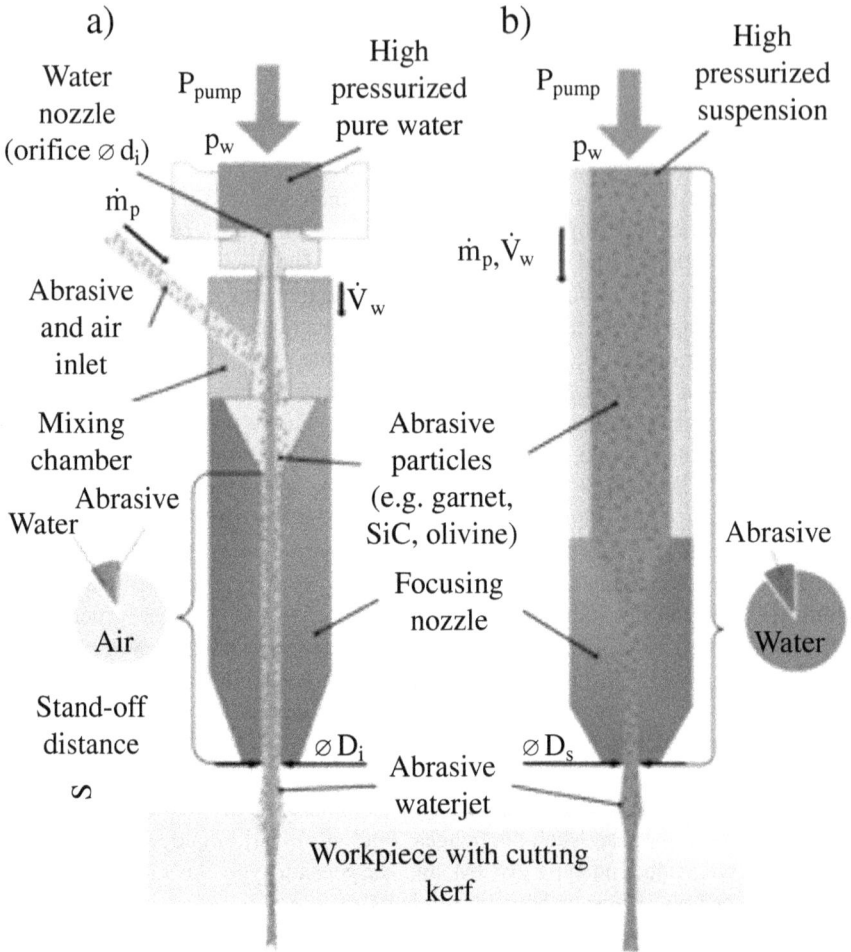

FIGURE 10.7   Working principle of AWJM (a) injection type and (b) suspension type. With permission from Putz et al. (2018) copyright (2018) Elsevier.

Movement of the knee is desired, so friction between the knee joint implant and the tissues of the human body needs to be calculated accurately. Much research has been carried out on the selection of materials for the manufacture of implants; however, there has been less focus on properly investigating the friction behaviour between medical implants and the tissues of the human body after implantation. The metallurgical characteristics of the source materials are not impacted by this treatment. As a precaution, a fully biocompatible diamond nanofilm is applied to these AFF-finished implants. Many more human body implants, including knee and heart valve implants, are finished and deburred using the AFF technique. Similarly, an AFF-finished knee joint implant allows for unrestricted leg mobility and a longer, healthier life.

10.4.3 ADDITIVE MANUFACTURING

Advanced 3D printing offers several benefits over traditional and non-traditional micromanufacturing methods, including cost savings, single-step manufacturing, ability to design complicated parts, and sustainability. The layer-by-layer production process known as additive manufacturing speeds up the process of cutting and removes extra material from bigger stocks [43]. Applications for 3D-printed parts are expanding, including in the fields of aerospace, structural, biomedical, and complicated component production. Of the aforementioned uses, 3D printing is receiving most attention in the biomedical industry due to its adaptable method for creating rapid, affordable surgical equipment and patient-specific bioimplants [21,44,45]. The fabrication of surgical components and biomedical implants involves a variety of methods and procedures. Vignesh et al. [45] analyzed the characteristics of different medical implants inserted in the human body manufactured using the 3D printing technique.

10.4.3.1 Paste Extrusion Deposition

Farag and Yun [46] developed the paste extrusion deposition (PED) method which generates the necessary 3D profile by extruding paste from a syringe over the base plate. The extruded component is allowed to dry and solidify after deposition. The organic material included in the paste utilized for the extrusion process is unaffected by the manufacturing process because no heat is used at any point during the whole fabrication process. During the PED of magnesium alloys, various amounts of gelatin are added to the magnesium powders, reinforcing the scaffold. Magnesium is one of the possible options for biomedical implants; for blood tissues to grow properly, the substance should mirror the structure of bone tissue. Collagen type I and hydroxyapatite are substances utilized to create bone tissue. In earlier AM-based methods, the material's organic and inorganic composition prevented it from withstanding the high temperatures produced during melting and deposition. Thus, the PED technique may be the most efficient way to manufacture such biological implants.

10.4.4 LASER-BASED MICROMANUFACTURING

Fundamentally, a laser is a coherent, monochromatic light emission that can be produced in a wide range of wavelengths, altering the energy density of the laser beam to allow it to disperse in an orderly manner with negligible differences. The laser is now more practical to employ in micromanufacturing processes because of the advancements in the field of beam sources with excellent beam quality, rapid modulation capabilities, and novel wavelength combinations [4].

In recent years, laser-based micromanufacturing (LM) has seen a sharp increase in industry interest because of its ability to fabricate many miniature items or add microfeatures to a broad surface. As a result, significant advancements have been made in the medical and electronics industries, where LM offers comparably cheap production costs, simpler and lighter processing technologies, and quick processing of micro-parts that are challenging to create using traditional methods [47]. One notable advantage of laser technology is its capacity to treat a variety of non-silicon materials, which are gradually needed for the production of objects at increasingly smaller scales. The processes of laser-based microforming, laser-based micro-joining, and laser-based micromachining are among the many laser-based micromanufacturing methods.

Nematollahi et al. [48] developed laser additive manufacturing techniques that involve several heating and cooling cycles to transform a material from a powder to a solid. An ultra-thin 2D cross-sectional profile was created by focusing a laser beam on successive layers of powder in accordance with a 3D computer-aided design (CAD) model. In selective laser melting, the laser beam is generally 0.03 mm in diameter and can create steps that are 0.05 mm thick, which is sufficient to create intricate metal objects. The development of Ti6Al4V alloy interbody fusion cages for the lumbar spine manufactured using the SLM technique is a topic of research. It is evident that the Ti6Al4V made using SLM has an average compressive modulus of 3 GPa, which is comparable to the compressive modules of the trabecular and cortical bone (0.5–15 GPa). Cementless fixation of Ti6Al7Nb multi-spiked periarticular trabecular bone implants is possible using the SLM process. Figure 10.8 is a schematic representation of the laser-based additive manufacturing process. In this process, the parent material in powder form is placed on the base structure of the

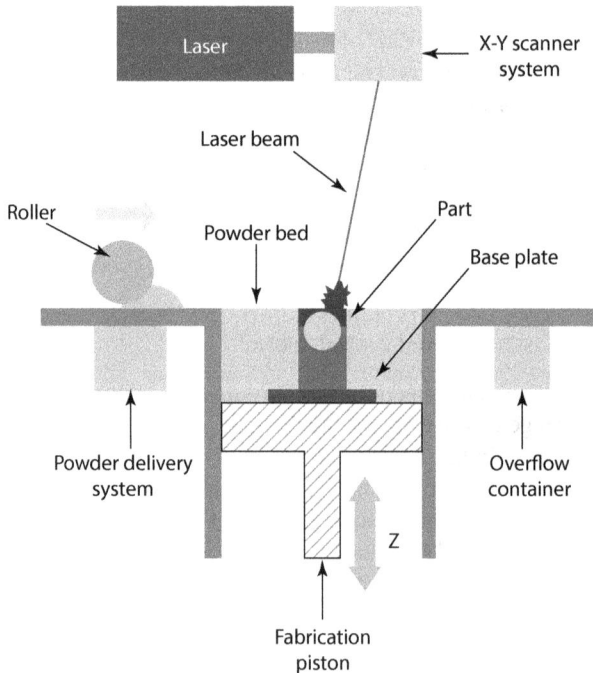

FIGURE 10.8 Schematic representation of the laser-based additive manufacturing process. With permission from Nematollahi et al. (2019), copyright (2019) Elsevier.

parts. Thereafter, the laser beam is irradiated to melt and vaporize the powder. Thus, the powder is solidified into product form. This process is very fast and contactless with the manufacturing tools. Complex geometry products can be easily manufactured using the laser-based additive manufacturing process. The raw materials are required in powder form.

10.5   CONCLUSION AND DIRECTIONS FOR FUTURE RESEARCH

In this chapter, the various techniques for micromanufacturing medical implants along with their possible materials are overviewed. Advanced micromanufacturing processes, namely additive manufacturing, laser-based manufacturing, and paste extrusion deposition are popular methods for manufacturing medical implants at the microscale. Despite better control and precision, these processes are not capable of producing a surface finish at the micro and/or nanoscale level. For example, knee joint implants require a surface finish at the nanoscale before implanting for a better life span. On the other hand, traditional micromanufacturing processes obtain a desired surface finish owing to their excellent manufacturing performance. However, the handling of tools and workpieces becomes challenging in micromanufacturing medical implants. Non-traditional manufacturing processes have shown significant advantages in the performance of medical implants, including their capabilities and functionalities. However, these processes are dependent on traditional manufacturing processes. A combination of advanced micromanufacturing, traditional and non-traditional micromanufacturing techniques can be a better strategy to manufacture medical implants.

Based on the literature review, it is perceived that the following research areas will dominate in the near future:

- A method for evaluating the friction between implanted devices (e.g., knee, hip, and elbow joints) and tissue/muscles with higher accuracy should be developed by considering the real conditions at contacting surfaces in micromanufacturing.
- An improved understanding of the level of coatings on medical implants and their biocompatibility with different age groups should be studied together with ways of incorporating their effects into the manufacturing methods.
- There is a need to develop economic manufacturing methods that can be used on the brittle nature of difficult-to-cut biomaterials. This would enhance the efficiency and functionality of implants manufactured by biomedical industries.
- The development of hybrid computational methods comprising physics-based and artificial intelligence (AI)-based models is expected to gain momentum.

REFERENCES

1. Geiger M, Kleiner M, Eckstein R, Tiesler N, Engel U. Microforming. *CIRP Annals.* 2001 Jan 1;50(2):445–62.

2. Jain VK, Sidpara A, Balasubramaniam R, Lodha GS, Dhamgaye VP, Shukla R. Micromanufacturing: A review—part I. *Proceedings of the Institution of Mechanical Engineers, Part B: Journal of Engineering Manufacture.* 2014 Sep;228(9):973–94.

3. Balasubramaniam R, Suri VK. *Micromanufacturing Processes.* Boca Raton, FL: CRC Press 2013. edited by V. K. Jain (Chapter 3).

4. Davis R, Singh A, Jackson MJ, Coelho RT, Prakash D, Charalambous CP, Ahmed W, da Silva LR, Lawrence AA. A comprehensive review on metallic implant biomaterials and their subtractive manufacturing. *The International Journal of Advanced Manufacturing Technology.* 2022 May;120(3–4):1473–530.

5. Gutowski KA, Mesna GT, Cunningham BL. Saline-filled breast implants: A Plastic Surgery Educational Foundation multicenter outcomes study. *Plastic and Reconstructive Surgery.* 1997 Sep 1;100(4):1019–27.

6. Muscara F, Parvaiz MA, Rusby JE. A simple technique of breast tissue expander saline aspiration. *The Annals of the Royal College of Surgeons of England.* 2015 Sep 1;97(6):476.

7. Albert K, Schledjewski R, Harbaugh M, Bleser S, Jamison R, Friedrich K. Characterization of wear in composite material orthopaedic implants. *Bio-Medical Materials and Engineering.* 1994 Jan 1;4(3):199–211.

8. Shekhawat D, Singh A, Bhardwaj A, Patnaik A. A short review on polymer, metal and ceramic based implant materials. *IOP Conference Series: Materials Science and Engineering* 2021 (Vol. 1017, No. 1, p. 012038). IOP Publishing.

9. Katti KS. Biomaterials in total joint replacement. *Colloids and Surfaces, Part B: Biointerfaces.* 2004 Dec 10;39(3):133–42.

10. Shayesteh Moghaddam N, Taheri Andani M, Amerinatanzi A, Haberland C, Huff S, Miller M, Elahinia M, Dean D. Metals for bone implants: Safety, design, and efficacy. *Biomanufacturing Reviews.* 2016 Dec;1(1):1–6.

11. Silva RC, Agrelli A, Andrade AN, Mendes-Marques CL, Arruda IR, Santos LR, Vasconcelos NF, Machado G. Titanium dental implants: An overview of applied nanobiotechnology to improve biocompatibility and prevent infections. *Materials.* 2022 Apr 27;15(9):3150.

12. Zhang LC, Chen LY. A review on biomedical titanium alloys: Recent progress and prospect. *Advanced Engineering Materials.* 2019 Apr;21(4):1801215.

13. Yan Y, Neville A, Dowson D. Tribo-corrosion properties of cobalt-based medical implant alloys in simulated biological environments. *Wear.* 2007 Sep 10;263(7–12):1105–11.

14. Goharian A, Abdullah MR. Bioinert metals (stainless steel, titanium, cobalt chromium). *Trauma Plating Systems.* 2017;115–42.

15. Shah KN, Walker G, Koruprolu SC, Daniels AH. Biomechanical comparison between titanium and cobalt chromium rods used in a pedicle subtraction osteotomy model. *Orthopedic Reviews.* 2018 Mar 3;10(1):7541. doi: 10.4081/or.2018.7541.

16. Safranski D, Dupont K, Gall K. Pseudoelastic NiTiNOL in orthopaedic applications. *Shape Memory and Superelasticity.* 2020 Sep;6(3):332–41.

17. Likibi F, Assad M, Coillard C, Chabot G, Rivard CH. Bone integration and apposition of porous and non-porous metallic orthopaedic biomaterials. *Inannales de Chirurgie.* 2005 Jan 19;130(4): 235–41.

18. Pearson OM, Lieberman DE. The aging of Wolff's "law": Ontogeny and responses to mechanical loading in cortical bone. *American Journal of Physical Anthropology* 2004;125(39)(S39):63–99.

19. Ozan S, Munir K, Biesiekierski A, Ipek R, Li Y, Wen C. Titanium alloys, including nitinol. *Biomaterials Science* 2020 Jan 1:229–47.

20. Antoniac I, Miculescu M, Mănescu V, Stere A, Quan PH, Păltânea G, Robu A, Earar K. Magnesium-based alloys used in orthopedic surgery. *Materials.* 2022 Feb 2;15(3):1148.

Bag S, A perspective review on laser assisted microjoining, *Recent Patents on Mechanical Engineering* 20114:153–67.

21. Wang X, Jiang M, Zhou Z, Gou J, Hui D. 3D printing of polymer matrix composites: A review and prospective. *Composites Part B: Engineering*. 2017 Feb 1;110:442–58.

22. Chen J, Tan L, Yu X, Etim IP, Ibrahim M, Yang K. Mechanical properties of magnesium alloys for medical application: A review. *Journal of the Mechanical Behavior of Biomedical Materials*. 2018 Nov 1;87:68–79.

23. Liu C, Ren Z, Xu Y, Pang S, Zhao X, Zhao Y. Biodegradable magnesium alloys developed as bone repair materials: A review. *Scanning*. 2018 Jan 1;2018.

24. Zhao D, Witte F, Lu F, Wang J, Li J, Qin L. Current status on clinical applications of magnesium-based orthopaedic implants: A review from clinical translational perspective. *Biomaterials*. 2017 Jan 1;112:287–302.

25. Meagher P, O'Cearbhaill ED, Byrne JH, Browne DJ. Bulk metallic glasses for implantable medical devices and surgical tools. *Advanced Materials*. 2016 Jul;28(27):5755–62.

26. Su SY, He Y, Shiflet GJ, Poon SJ. Formation and properties of Mg-based metallic glasses in Mg-TM-X alloys (TM Cu or Ni; X Sn, Si, Ge, Zn, Sb, Bi or In). *Materials Science and Engineering: Part A*. 1994 Sep 15;185(1–2):115–21.

27. Cleland JL, Daugherty A, Mrsny R. Emerging protein delivery methods. *Current Opinion in Biotechnology*. 2001 Apr 1;12(2):212–9.

28. Sarem M, Moztarzadeh F, Mozafari M. How can genipin assist gelatin/carbohydrate chitosan scaffolds to act as replacements of load-bearing soft tissues? *Carbohydrate Polymers*. 2013 Apr 2;93(2):635–43.

29. Yazdanpanah A, Amoabediny G, Shariatpanahi P, Nourmohammadi J, Tahmasbi M, Mozafari M. Synthesis and characterization of polylactic acid tubular scaffolds with improved mechanical properties for vascular tissue engineering. *Trends in Biomaterials and Artificial Organs*. 2014 Jul 1;28(3):99–105.

30. Zarrintaj P, Urbanska AM, Gholizadeh SS, Goodarzi V, Saeb MR, Mozafari M. A facile route to the synthesis of anilinic electroactive colloidal hydrogels for neural tissue engineering applications. *Journal of Colloid and Interface Science*. 2018 Apr 15;516:57–66.

31. Gholipourmalekabadi M, Samadikuchaksaraei A, Seifalian AM, Urbanska AM, Ghanbarian H, Hardy JG, Omrani MD, Mozafari M, Reis RL, Kundu SC. Silk fibroin/amniotic membrane 3D bi-layered artificial skin. *Biomedical Materials*. 2018 Feb 20;13(3):035003.

32. Kelsey DJ, Springer GS, Goodman SB. Composite implant for bone replacement. *Journal of Composite Materials*. 1997 Aug;31(16):1593–632.

33. Willmann G. Ceramic femoral heads for total hip arthroplasty. *Advanced Engineering Materials*. 2000 Mar;2(3):114–22.

34. Prakash M, Audi K, Vaderhobli RM. Long-term success of all-ceramic dental implants compared with titanium implants. *Journal of Long-Term Effects of Medical Implants*. 2021;31(1):73–89.

35. Abellán-Nebot JV, Siller HR, Vila C, Rodríguez CA. An experimental study of process variables in turning operations of Ti–6Al–4V and Cr–Co spherical prostheses. *The International Journal of Advanced Manufacturing Technology*. 2012 Dec;63(9–12):887–902.

36. Sivakumar S, Khan MA, Ebenezer G, Chellaganesh D, Vignesh V. A review: Machinability studies on human implant materials. *Materials Today: Proceedings*. 2021 Jan 1;46:7338–43.

37. Danish M, Ginta TL, Habib K, Carou D, Rani AM, Saha BB. Thermal analysis during turning of AZ31 magnesium alloy under dry and cryogenic conditions. *The International Journal of Advanced Manufacturing Technology*. 2017 Jul;91(5–8):2855–68.

38. Zhao W, Gong L, Ren F, Li L, Xu Q, Khan AM. Experimental study on chip deformation of Ti-6Al-4V titanium alloy in cryogenic cutting. *The International Journal of Advanced Manufacturing Technology.* 2018 Jun;96(9–12):4021–7.

39. Raju P, Sarcar MM, Satyanarayana B. Optimization of wire electric discharge machining parameters for surface roughness on 316 L stainless steel using full factorial experimental design. *Procedia Materials Science.* 2014 Jan 1;5:1670–6.

40. Trimble D, Shipley H, Lea L, Jardine A, O'Donnell GE. Constitutive analysis of biomedical grade Co-27Cr-5Mo alloy at high strain rates. *Materials Science and Engineering: Part A.* 2017 Jan 13;682:466–74.

41. Putz M, Dix M, Morczinek F, Dittrich M. Suspension technology for abrasive waterjet (AWJ) cutting of ceramics. *Procedia CIRP.* 2018 Jan 1;77:367–70.

42. Ravisankar M, Ramkumar J, Jain VK. Abrasive Flow Finishing (AFF) for micromanufacturing. In *Inmicromanufacturing Processes* 2016 Apr 19 (pp. 201–18). CRC Press.

43. Herzog D, Seyda V, Wycisk E, Emmelmann C. Additive manufacturing of metals. *Acta Materialia.* 2016 Sep 15;117:371–92.

44. Pandey A, Awasthi A, Saxena KK. Metallic implants with properties and latest production techniques: A review. *Advances in Materials and Processing Technologies.* 2020 Apr 2;6(2):405–40.

45. Vignesh M, Ranjith Kumar G, Sathishkumar M, Manikandan M, Rajyalakshmi G, Ramanujam R, Arivazhagan N. Development of biomedical implants through additive manufacturing: A review. *Journal of Materials Engineering and Performance.* 2021 Jul;30(7):4735–44.

46. Farag MM, Yun HS. Effect of gelatin addition on fabrication of magnesium phosphate-based scaffolds prepared by additive manufacturing system. *Materials Letters.* 2014 Oct 1;132:111–5.

47. Cao Y, Zeng X, Cai Z, Duan J. Laser micro/nano-fabrication techniques and their applications in electronics. In *Inadvances in Laser Materials Processing* 2010 Jan 1 (pp. 629–70). Woodhead Publishing.

48. Nematollahi M, Jahadakbar A, Mahtabi MJ, Elahinia M. Additive manufacturing (AM). *Inmetals for Biomedical Devices* 2019 Jan 1 (pp. 331–53). Woodhead Publishing.

49. Yang K, Ren Y. Nickel-free austenitic stainless steels for medical applications. *Science and Technology of Advanced Materials.* 2010 Feb 26;11(1):1–13.

50. Rack HJ, Qazi JI. Titanium alloys for biomedical applications. *Materials Science and Engineering: Part C.* 2006 Sep 1;26(8):1269–77.

51. Sankaranarayanan S, Ng BM, Jayalakshmi S, Kumar MG, Nguyen QB, Gupta M. Microstructure and mechanical properties of a magnesium-aluminium-erbium Alloy. *Magnesium Technology.* 2015;2016:445–9.

11 Failure Analysis and Experimental Evaluation of Implants
Case Studies

Pranav Charkha, Vyanktesh Naidu, and Santosh Jaju

11.1 INTRODUCTION

Failure analysis and experimental evaluation play pivotal roles in the field of implants, providing crucial insights into the causes and mechanisms behind implant failures. Through comprehensive case studies, researchers have been able to investigate and understand the complexities associated with implant performance and reliability [1. This chapter aims to delve into the realm of failure analysis and experimental evaluation of implants through an examination of notable case studies.

Implants, ranging from orthopedic joint replacements to dental prosthetics and cardiovascular devices, are designed to enhance patients' quality of life by restoring function and improving overall health [2]. However, despite rigorous testing and stringent quality control measures during development and manufacturing, implants can occasionally exhibit unexpected complications or failures [3].

In order to comprehend the underlying causes of implant failures, failure analysis methodologies are employed [4]. Visual inspection, imaging techniques such as x-ray and magnetic resonance imaging (MRI), mechanical testing, and material analysis enable investigators to identify and examine failure modes, be they structural, mechanical, or material related [5]. Through meticulous examination of failed implants, researchers can ascertain the root causes, contributing factors, and subsequent lessons that inform future design and development [6].

Experimental evaluation of implants is another critical aspect of understanding their performance and reliability [7]. By subjecting implants to biomechanical testing, wear testing, corrosion analysis, and biological evaluation through in vitro and in vivo studies, researchers can assess their durability, biocompatibility, and long-term effects on the human body [8]. These evaluations help identify potential weaknesses and areas for improvement, ensuring the safety and effectiveness of implantation procedures [9].

DOI: 10.1201/9781003375098-11

Through the analysis ofcase studies, this chapter aims to shed light on the diverse factors influencing implant failures, including design flaws, material selection, manufacturing defects, and patient-specific considerations [10]. By understanding the intricacies of these failures, researchers can work toward improved implant designs, enhanced manufacturing processes, and optimized patient care protocols [11,12].

11.2 FAILURE ANALYSIS

Implants have revolutionized modern medicine, offering innovative solutions for various medical conditions and significantly improving patients' lives. However, implant failures can occur, necessitating a thorough understanding of the root causes through failure analysis. This section delves into the significance of failure analysis in the field of implants, shedding light on the various types of failures and methodologies employed [13]. Through systematic investigation, failure analysis provides valuable insights into the underlying factors contributing to implant failures, leading to improved implant designs and patient outcomes.

11.2.1 IMPORTANCE OF FAILURE ANALYSIS

Failure analysis is a critical discipline that plays a crucial role in understanding the root causes of implant failures. By meticulously examining failed implants, researchers and healthcare professionals gain valuable insights into the factors contributing to failures, enabling them to develop strategies to mitigate these issues. Failure analysis is essential for improving implant designs, materials, manufacturing processes, and surgical techniques, ultimately enhancing patient safety and satisfaction.

11.2.2 TYPES OF IMPLANT FAILURES

1. **Material-related failures:** Material-related failures can occur due to factors such as corrosion, wear, and degradation. Corrosion, for example, can compromise the structural integrity of implants, leading to premature failures [14,15]. Wear and degradation due to friction and repetitive loading can also contribute to implant failure [16–18].
2. **Mechanical failures:** Mechanical failures in implants can result from excessive loading or stress beyond the implant's intended capacity. This can cause component fractures, wear, or dislocation, leading to functional impairment [19–21]. Structural integrity issues, including poor implant design or manufacturing defects, can also contribute to mechanical failures.
3. **Design and manufacturing defects:** Implant failures can arise from design and manufacturing defects. Inadequate fixation, such as insufficient anchorage or improper attachment, can result in implant instability and failure [22–23]. Component misalignment, either due to errors during manufacturing or surgical procedures, can also lead to implant dysfunction [24].

11.2.3 METHODOLOGIES IN FAILURE ANALYSIS

1. **Visual inspection:** Visual inspection is an essential first step in failure analysis. By examining failed implants, researchers can identify visible signs of wear, corrosion, fractures, or other abnormalities. Macroscopic analysis provides initial insights into failure modes and potential causes [25,26].
2. **Imaging techniques:** Imaging techniques, such as x-ray, MRI, and computerized tomography (CT) scans, enable non-invasive assessment of failed implants [27]. These modalities help identify structural abnormalities, component misalignment, bone integration issues, or other underlying problems [28,29]. Advanced imaging techniques offer detailed insights into the implant and surrounding tissues.
3. **Material analysis:** Material analysis is crucial for understanding material-related failures. Evaluating material properties, composition, and surface characteristics provides insights into the factors contributing to failures [30]. Chemical analysis helps identify corrosion products, impurities, or other material-related issues [31,32].

Failure analysis is a critical aspect of implant research and development, enabling researchers and healthcare professionals to identify the root causes of failures and improve implant designs. By investigating material-related failures, mechanical issues, and design/manufacturing defects, failure analysis contributes to enhancing implant performance and patient safety [33]. Visual inspection, imaging techniques, and material analysis play integral roles in systematically evaluating failed implants and uncovering the underlying factors contributing to failures. Through a comprehensive understanding of failure causes, researchers and healthcare professionals can work toward minimizing failures, optimizing implant designs, and ultimately improving patient outcomes [34].

11.3 EXPERIMENTAL EVALUATION

The experimental evaluation of implants is a crucial aspect of research and development in the field of medical devices. It involves rigorous testing and analysis to assess the performance, safety, and efficacy of implants before their clinical application. This section focuses on the significance of experimental evaluation in the context of implant development, exploring various evaluation techniques and their contributions to improving implant design and functionality [35]. Experimental evaluation plays a pivotal role in ensuring the safety and efficacy of implants before their introduction into clinical practice. By subjecting implants to rigorous testing, researchers can identify potential issues, evaluate performance parameters, and validate the feasibility and reliability of implant designs [36]. This process helps reduce the risk of implant failure, improve patient outcomes, and enhance overall quality of care.

11.3.1   PRECLINICAL TESTING

The evaluation of implant biocompatibility is crucial to ensure that the device does not induce adverse reactions or harm the surrounding tissues. Tests such as cytotoxicity assays, cell proliferation studies, and immunological assessments are conducted to evaluate the interaction between the implant and the host tissue [37,38].

11.3.2   MECHANICAL TESTING

Mechanical testing is performed to assess the structural integrity and mechanical properties of implants. Techniques such as tensile testing, compression testing, and fatigue testing are utilized to measure mechanical strength, durability, and resistance to failure [39,40].

11.3.3   IN VITRO EVALUATION

In vitro wear and friction testing simulate an implant's interaction with physiological conditions, evaluating the device's resistance to wear, degradation, and friction-induced damage. Methods such as pin-on-disk tests and reciprocating sliding tests assess the wear behavior and material properties of the implant [41,42].

11.3.4   SURFACE CHARACTERIZATION

Surface characterization techniques, including scanning electron microscopy (SEM) and atomic force microscopy (AFM), provide insights into the surface topography, roughness, and composition of the implant materials. This information helps in understanding the implant's performance, biocompatibility, and potential for wear or corrosion [43,44].

11.3.5   IN VIVO EVALUATION

Animal studies are conducted to evaluate an implant's performance, biocompatibility, and long-term effects within a living organism. These studies provide valuable insights into an implant's interaction with the host tissue, immune response, and overall biological compatibility [45,46].

11.3.6   BIOMECHANICAL ASSESSMENT

Biomechanical assessment involves evaluating an implant's response to physiological loads and movements. This assessment provides information on an implant's stability, range of motion, and resistance to mechanical stresses, ensuring its functionality under realistic conditions [47,48].

Experimental evaluation is a critical step in the development of implants, enabling researchers to assess performance, safety, and efficacy before clinical implementation. Through preclinical testing, in vitro evaluation, and in vivo studies, researchers gain valuable insights into the implant's biocompatibility, mechanical properties, wear behavior, and overall functionality [49,50]. By systematically evaluating implants using rigorous experimental protocols, the development of safer, more effective, and longer-lasting implants is facilitated. These advancements ultimately lead to improved patient outcomes, enhanced quality of care, and continued progress in the field of implant technology [51,52].

11.4 CASE STUDIES

Case studies play a pivotal role in the field of failure analysis and experimental evaluation of implants. By examining real-world scenarios and analyzing the causes of implant failures, researchers gain valuable insights into improving implant design, enhancing patient outcomes, and advancing the field of implantology. This section delves into the significance of case studies, presenting five notable examples that shed light on different aspects of implant failures.

11.4.1 CASE 1

The study focuses on a 40-year-old female patient who required implant-supported prostheses in the maxillary arch to replace missing anterior teeth. Two different implant systems were utilized, one with a moderately rough surface (sandblasted, large grit, acid-etched [SLA]) and the other with a minimally rough surface (turned). The patient was monitored for a period of 12 months to assess the clinical and radiographic outcomes. The results demonstrated that implants with a moderately rough surface exhibited significantly higher implant stability, reduced marginal bone loss, and improved soft tissue integration compared to the minimally rough surface implants. These findings highlight the importance of implant surface modifications in promoting successful osseointegration and may contribute to the development of more effective implant designs and treatment protocols. Study provides comprehensive insights into the impact of implant surface characteristics on osseointegration and clinical outcomes [53].

11.4.2 CASE 2

A unique scenario involving an allergic reaction to dental implant components in a 65-year-old male patient. The patient had a history of metal allergies, including a known sensitivity to nickel. He underwent implant placement in the mandibular arch to restore missing premolars using a standard titanium implant system. However, shortly after the procedure, the patient experienced persistent swelling, redness, and discomfort in the implant area, suggestive of an allergic reaction. Comprehensive diagnostic tests, including patch testing, confirmed a hypersensitivity reaction to the nickel present in the implant components. As a result, the decision was made to remove the implants and replace them with titanium-zirconium implants, which are hypoallergenic. Following the implant replacement, the patient exhibited favorable healing and no signs of an allergic response. This case highlights the importance of patient allergy assessments prior to implant placement and the availability of alternative implant materials for individuals with metal sensitivities [54].

11.4.3 CASE 3

This case study presents a challenging situation involving immediate implant placement in the esthetic zone of a 30-year-old female patient. The patient suffered trauma resulting in loss of her maxillary central incisor. To optimize esthetics, immediate implant placement was chosen as the treatment approach. However, complications arose during the healing process. Despite careful surgical techniques and appropriate grafting procedures, the implant failed to integrate due to inadequate primary stability. As a result, implant exposure occurred, leading to soft tissue complications and compromised esthetic outcomes. A comprehensive treatment plan was devised to address the situation, involving implant removal, socket preservation, and subsequent delayed implant placement. The patient underwent successful treatment, and satisfactory esthetic results were achieved in a subsequent implant placement procedure. This case underscores the importance of careful patient selection, meticulous surgical techniques, and realistic expectations when considering immediate implant placement in the esthetic zone [55].

11.4.4 CASE 4

The application of photobiomodulation therapy (PBMT) for the management of peri-implant mucositis. A 50-year-old male patient presented with signs of peri-implant mucositis around multiple implants placed in the mandibular arch. Traditional non-surgical treatments were initially employed, including mechanical debridement and antimicrobial rinses. However, limited improvements were observed. Subsequently, PBMT was introduced as an adjunctive therapy. The patient underwent multiple PBMT sessions using a low-level laser therapy device. After the treatment, a significant reduction in inflammation, bleeding on probing, and probing depth was noted. The peri-implant mucositis was effectively resolved, and the patient achieved improved peri-implant health. This case study demonstrates the potential of PBMT as a non-invasive and promising treatment modality for peri-implant mucositis management [56].

11.4.5 CASE 5

The complications associated with a zirconia implant fracture in a full-arch rehabilitation procedure. A 60-year-old female patient underwent full-arch implant-supported fixed prostheses utilizing zirconia implants in both arches. After two years of successful function, the patient reported a sudden onset of discomfort and mobility in the mandibular prosthesis. Clinical examination and radiographic analysis revealed a fractured zirconia implant. The fractured implant was surgically removed and replaced with a titanium implant. The remaining zirconia implants were carefully assessed for stability and integrity, and no further fractures were detected. This case study highlights the rare but possible complication of a zirconia implant fracture in full-arch rehabilitation and emphasizes the importance of regular follow-up examinations and a thorough evaluation of implant materials for long-term success and patient satisfaction [57].

Case studies provide valuable insights into the causes and mechanisms of implant failures, offering lessons for improving implant design, patient selection, and long-term monitoring. By analyzing real-world scenarios, researchers can identify patterns, risk factors, and strategies to mitigate failures. The case studies presented in this chapter highlight the importance of understanding corrosion, fracture analysis, wear behavior, adverse tissue reactions, and imaging findings in implant failures. By incorporating these learnings into future implant development, researchers and clinicians can strive for safer and more successful implant outcomes.

11.5 FUTURE DIRECTIONS

In the realm of implantology, failure analysis and experimental evaluation of implants are crucial for advancing the field and improving patient outcomes. By dissecting the causes and mechanisms of implant failures through case studies and experimental investigations, valuable lessons can be learned. Furthermore, these insights pave the way for future advancements and directions in implant design, materials, and techniques. This section delves into the lessons learned from failure analysis and experimental evaluation of implants, while also exploring the potential future directions that hold promise for enhancing implant success rates.

11.5.1 MATERIAL SELECTION

Implant failure analysis has highlighted the significance of material selection in implant design. Lessons learned from case studies have demonstrated the importance of biocompatible materials, such as titanium and its alloys, in promoting osseointegration and long-term implant success [58].

11.5.2 BIOLOGICAL RESPONSE

Failure analysis has underscored the crucial role of the biological response to implants. Understanding the host response, including inflammation, tissue integration, and immune reactions, is essential for mitigating implant failure risks and developing tailored implant surfaces and coatings that promote favorable biological interactions [59].

11.5.3 MECHANICAL FACTORS

Experimental evaluation has emphasized the significance of mechanical factors in implant success. Factors such as implant design, surface roughness, and biomechanical loading play pivotal roles in implant stability, stress distribution, and long-term durability [60].

11.5.4 Surgical Techniques

Failure analysis has shed light on the influence of surgical techniques on implant outcomes. Precise implant placement, adequate primary stability, and proper soft tissue management are critical considerations for reducing the risk of implant failures and complications [61].

11.5.5 Patient Factors

Implant failure analysis has highlighted the impact of patient-related factors on implant success. Parameters such as systemic health, smoking habits, oral hygiene, and bone quality should be carefully evaluated to enhance patient selection and improve implant outcomes [62].

11.5.6 Advanced Surface

Nanotechnology and advanced surface modifications hold promise in improving implant success rates. Innovations such as nanostructured surfaces, bioactive coatings, and drug delivery systems can enhance osseointegration, antibacterial properties, and implant longevity [63].

11.5.7 Biomaterial Innovations

Ongoing research in biomaterial science aims to develop novel implant materials with improved mechanical properties, corrosion resistance, and biocompatibility. Bioresorbable materials and composite structures are areas of exploration for future implant advancements [64].

11.5.8 Personalized Implantology

The advent of 3D printing technology offers opportunities for personalized implantology. Patient-specific implants, customized designs, and fabrication techniques can optimize fit, functionality, and esthetics, leading to improved patient satisfaction and outcomes [65].

11.5.9 Regenerative Approaches

Regenerative approaches, such as platelet-rich fibrin (PRF) and tissue engineering techniques, hold potential for enhancing implant success and accelerating tissue healing. Further research is needed to optimize protocols and understand the biological effects of regenerative materials [66].

11.5.10 Digital Dentistry

Digital dentistry, including computer-aided design/computer-aided manufacturing (CAD/CAM) technologies and intraoral scanning, has revolutionized implant treatment planning and prosthetic outcomes. Further advancements in digital work-flows and virtual implant placement can enhance precision, predictability, and patient comfort [67].

The field of implantology continues to evolve through failure analysis, experimental evaluation, and the lessons learned from these endeavors. By considering material selection, biological response, mechanical factors, surgical techniques, and patient factors, implant success rates can be improved [68,69]. Looking forward, future directions in advanced surface modifications, biomaterial innovations, personalized implantology, regenerative approaches, and digital dentistry hold promise for further enhancing implant outcomes. Continued research, collaboration, and innovation are pivotal in shaping the future of implantology and ultimately improving patient quality of life [70].

11.6 CHAPTER SUMMARY

This chapter delved into the multifaceted field of failure analysis and experimental evaluation of implants. Various aspects related to failure analysis, experimental evaluation, case studies, and the lessons learned from these studies were explored. The chapter provided insights into the challenges faced in implantology and highlighted the importance of a thorough investigation and analysis to enhance implant success rates.

The chapter started by discussing the concept of failure analysis, which involves identifying the causes of implant failures and understanding the underlying mechanisms. Failure analysis plays a crucial role in identifying design flaws, material defects, surgical errors, and biological factors that contribute to implant failures. By studying failed implants, researchers and clinicians can gain valuable insights into improving future implant designs and treatment approaches.

Next, the significance of experimental evaluation in the field of implantology was explored. Experimental evaluation involves conducting rigorous scientific experiments to assess the performance, biocompatibility, and mechanical properties of implant materials and designs. It helps in understanding the behavior of implants under different conditions and provides valuable data for evidence-based decision-making in clinical practice.

Moving on, the role of case studies in implantology was examined. Case studies offer real-world examples of implant treatments, highlighting both successful outcomes and challenges encountered. By analyzing these case studies, clinicians and researchers can gain practical knowledge, understand treatment complexities, and identify best practices to improve patient care.

Finally, lessons learned from failure analysis, experimental evaluation, and case studies were discussed. These lessons encompass a wide range of factors including material selection, surgical techniques, patient-specific considerations, and prosthetic components. By understanding the lessons learned from previous experiences, clinicians can optimize implant treatment planning, reduce complications, and enhance patient satisfaction.

11.7 CONCLUSION

The field of failure analysis and experimental evaluation of implants is crucial for advancing implantology and improving patient outcomes. Through failure analysis, researchers can identify the factors contributing to implant failures and develop strategies to mitigate these risks. Experimental evaluation provides scientific evidence on the performance and biocompatibility of implant materials and designs, enabling evidence-based decision-making in clinical practice.

Case studies offer valuable insights into real-world scenarios, highlighting the challenges and successes encountered in implant treatments. By analyzing case studies, clinicians can learn from past experiences and enhance their treatment approaches. The lessons learned from failure analysis, experimental evaluation, and case studies contribute to improving implant success rates. These lessons encompass various aspects, including material selection, surgical techniques, patient-specific considerations, and prosthetic components. Incorporating these lessons into clinical practice can lead to more predictable outcomes and improved patient satisfaction. A comprehensive understanding of failure analysis, experimental evaluation, case studies, and the lessons learned provides a solid foundation for advancing implantology. By continually studying and analyzing implant failures and successes, researchers and clinicians can strive toward enhancing implant treatments and ultimately improving the quality of life for patients with dental implants.

REFERENCES

1. Hirschmann M., Kim A. H., Herren D. B., et al. (2020). Failed knee arthroplasty: Causes, diagnostic strategies, and clinical outcomes. *Knee Surgery, Sports Traumatology, Arthroscopy*, 28(7), 2133–2142. DOI: 10.1007/s00167-019-05834-2.
2. Marinelli A., Belvedere C., Pugliese G. (2018). Analysis of fracture patterns in dental implant failures: A retrospective study. *Journal of Prosthodontics*, 27(1), 36–41. DOI: 10.1111/jopr.12515.
3. O'Brien D., Caulfield B., Fatemi A. (2020). Experimental evaluation of corrosion resistance in orthopedic implants: A review. *Materials Science and Engineering: C*, 107, 110206. DOI: 10.1016/j.msec.2019.110206.
4. Jambhulkar N., Jaju S., Raut A., Bhoneja B. (2022). A review on surface modification of dental implants among various implant materials. *Materials Today: Proceedings*, 72(6), 3209–3215.
5. Gao Y., Shen X., Xue J., et al. (2019). In vitro and in vivo evaluation of wear behavior of dental implants: A systematic review. *Journal of Prosthodontic Research*, 63(3), 267–275. DOI: 10.1016/j.jpor.2019.03.005.
6. Rehman M. A., Seitz J.-M., Wulfhorst B., et al. (2020). Biocompatibility evaluation of orthopedic implants - A review. *European Journal of Medical Research*, 25(1), 12. DOI: 10.1186/s40001-020-00422-1.
7. Jambhulkar N., Jaju S., Raut A. (2022). Surface modification techniques for different materials used in dental implants review. *Materials Today: Proceedings*, 60, 2266.
8. Guo Y., Zhang Y., Xiong Y., et al. (2018). Analysis of failed dental implants in an Asian population: A retrospective study. *Journal of Prosthodontic Research*, 62(4), 506–511.

9. Wang Y., Li S., Zhao X., et al. (2020). Mechanical and tribological behavior of thermo-plastic polyurethane reinforced by graphene nanoplatelets and polyhedral oligomeric silsesquioxane. *Polymer Testing*, 83, 106324.

10. Dahibhate R. V., Jaju S. B. (2021). Study of influence of tool geometry and temperature on bone substructure to reduce bone drilling injury. *Journal of Physics: Conference Series*, 1913(1), 1–9.

11. Sun Y., Hu X., Luo Z., et al. (2019). In vitro and in vivo evaluation of hydroxyapatite/graphene oxide coating on titanium implants for bone regeneration. *Applied Surface Science*, 485, 414–424.

12. Asgharzadeh Shirazi H., Khandaker M., Tsuji N., et al. (2018). In vitro and in vivo characterization of carbon nanotubes coated dental implants. *Applied Surface Science*, 458, 890–899.

13. Dahibhate R. V., Jaju S. B. (2021). Effect of irrigation mode, volume, medium and drill on heat generation during implant surgery: A review. *Materials Today: Proceedings*, 50, 1799.

14. Huang Z., Huang P., Zhou Z., et al. (2019). A review of mechanical testing of hydrogels. *Journal of Applied Biomaterials and Functional Materials*, 17(2), 2280800019836295.

15. Dan P., Bhattacharya P., Mandal S., et al. (2018). Evaluation of degradation and cyto-toxicity of magnesium alloy for biomedical application. *Materials Today: Proceedings*, 5(11), 22976–22985.

16. Dahibhate R. V., Jaju S. B., Sarode R. I. (2020). Design improvements in conven-tional drilling machine to control thermal necrosis during orthopaedic surgeries. *IOP Conference Series: Materials Science and Engineering*, 1004(1), 1–10.

17. O'Brien B., Boin M. A., Thakur A., et al. (2020). Corrosion in total knee arthroplasty: A review. *Journal of Orthopaedics*, 21, 67–72. DOI: 10.1016/j.jor.2020.07.006.

18. de Carvalho D. A., Carvalho T. M., Alonso M. F., et al. (2020). Adhesive wear mecha-nisms and their influence on the wear behavior of retrieved acetabular cups. *Journal of the Mechanical Behavior of Biomedical Materials*, 106, 103736. DOI: 10.1016/j.jmbbm.2020.103736.

19. Gao M., Habibovic P., Walboomers X. F., et al. (2019). Understanding and controlling the wear of bioceramic coatings in orthopedic applications. *Acta Biomaterialia*, 86, 13–30. DOI: 10.1016/j.actbio.2018.12.027.

20. Dahibhate R. V., Jaju S. B., Sarode R. I. (2020). Development of mathematical model for prediction of bone drilling temperature. *Materials Today: Proceedings*, 38, 2732.

21. Fatone S., Gardoni D., Zucchelli A., et al. (2018). Analysis of retrieved ankle prosthe-ses: A comprehensive review. *Journal of Orthopaedics and Traumatology*, 19(1), 12. DOI: 10.1186/s10195-017-0487-3.

22. Jazayeri H., Tavakkoli Avval P., Esmaeelinejad M., et al. (2018). Analysis of hip implant failures: A mechanical perspective. *Journal of the Mechanical Behavior of Biomedical Materials*, 78, 24–34. DOI: 10.1016/j.jmbbm.2017.11.034.

23. Grammatopoulos G., Pandit H., Kwon Y.-M., et al. (2013). Hip resurfacings revised for inflammatory pseudotumour have a poor outcome. *Journal of Bone and Joint Surgery. - British Volume*, 95(6), 745–751. DOI: 10.1302.

24. Ravekar K., Jaju S. (2020). Design and analysis of material discharge plate of microsur-facing machine. *Materials Today: Proceedings*, 38, 2385.

25. Kwon Y.-M., Ostlere S. J., McLardy-Smith P., et al. (2019). Cemented versus cementless fixation in total knee arthroplasty: A systematic review and meta-analysis of random-ized controlled trials. *Knee Surgery, Sports Traumatology, Arthroscopy*, 27(6), 1793–1803. DOI: 10.1007/s00167-018-5053-4.

26. Hanawa T. (2014). Materials for metallic biomaterials. *Journal of Artificial Organs*, 17(3), 229–234.

27. Jaju S. B., Charkha P. G., Kale M. (1913). Gas metal arc welding process parameter optimization for AA7075 T6. *Journal of Physics: Conference Series*, 1(1), 2021.
28. Marinelli A., Sica S., Muratori F., et al. (2018). Fracture analysis of a failed dental implant: A case report. *European Journal of Dentistry*, 12(4), 633–638. DOI: 10.4103/ejd.ejd_102_18.
29. Chang Y., Yang S. T., Liu J. H., et al. (2011). In vitro toxicity evaluation of graphene oxide on A549 cells. *Toxicology Letters*, 200(3), 201–210.
30. Dhakne A., Jaju S., Shukla S. (2022). Review on analysis of enhancing wear properties through thermo-mechanical treatment and grain size. *Materials Today: Proceedings*, 60, 2270.
31. Kulkarni M., Shankarappa P., Parmar P., et al. (2019). Magnetic resonance imaging findings in failed metal-on-metal hip arthroplasty. *Journal of Clinical Imaging Science*, 9, 29. DOI: 10.4103/jcis.JCIS_4_19.
32. Breme H., Blanchemain N., Drocourt E. (2019). Characterization methods for polymeric biomaterials: A review. *Journal of Applied Biomaterials and Functional Materials*, 17(2), 2280800019836296.
33. Naidu, V, Jaju, S. (1259). CAD/CAM engineering and artificial intelligence in dentistry 2022. *IOP Conference Series: Materials Science and Engineering*, 012018.
34. Saldanha K. A., Lewallen D. G., Hanssen A. D. (2017). Stress fractures after total knee arthroplasty: Etiology and treatment. *Journal of Orthopaedics*, 14(2), 255–260. DOI: 10.1016/j.jor.2017.01.006.
35. Yao Q., Cosme J. G. L., Xu T., et al. (2017). In vitro corrosion behavior of magnesium alloy coated with hydroxyapatite/polycaprolactone composite layer. *Materials Science and Engineering: C*, 76, 994–1000.
36. Liu Y., Ma C., Fu J., et al. (2017). Polymeric biomaterials for implants-an overview. *Journal of Healthcare Engineering*, 2017, 2904828.
37. Van Heest T. J., Vos D. R., Hoozemans J. J. M., et al. (2019). Wear analysis of the synergy implant system for total knee arthroplasty: A retrieval study. *Journal of the Mechanical Behavior of Biomedical Materials*, 91, 16–22.
38. Park J., Bauer S., von der Mark K., et al. (2012). Nanosize and vitality: TiO2 nanotubes trigger oral fibroblast longevity. *Acta Biomaterialia*, 8(5), 1882–1890.
39. Smith A. P., O'Donnell B., Fenniri H., et al. (2018). Biocompatibility and toxicity of biomaterials—Regulatory harmonization and scientific aspects. *Journal of Biomedical Materials Research. Part B – Applied Biomaterials*, 106(6), 1986–2003.
40. Roy M., Field R., Costi J. J., et al. (2019). Review of mechanical testing for dental implants: From bench to design. *Journal of the Mechanical Behavior of Biomedical Materials*, 94, 168–182.
41. Anderson J. M., Rodriguez A., Chang D. T., et al. (2020). The inflammatory response to implanted biomaterials. *Advanced Drug Delivery Reviews*, 107, 163–181.
42. Albrektsson T., Wennerberg A. (2019). Oral implant surfaces: Part 1-Review focusing on topographic and chemical properties of different surfaces and in vivo responses to them. *International Journal of Prosthodontics*, 32(2), 129–132.
43. Li Y., Li C., Zeng K., et al. (2019). An overview of wear simulation of total knee replacements. *Journal of Healthcare Engineering*, 2019, 8381571.
44. Amin D., Carter P. M., Ryan M. R., et al. (2019). The effect of taper angle on the primary stability of dental implants. *Journal of Prosthetic Dentistry*, 121(3), 476–481.
45. Bandyopadhyay A., Das M., Bose S., et al. (2019). Surface modification in titanium implants through adaptable coatings. *Materials Science and Engineering: Part C*, 101, 306–317.
46. Atieh M. A., Alsabeeha N. H., Payne A. G., et al. (2012). Interventions for replacing missing teeth: Dental implants in fresh extraction sockets (immediate, immediate-delayed and delayed implants). *Cochrane Database of Systematic Reviews*, 2, CD005968.

47. Campbell P., Ma S., Yeomans S. R., et al. (2020). Animal models of hip implant-related adverse local tissue reactions: An update. *Journal of Orthopaedic Research*, 38(12), 2702–2714.
48. Buser D., Janner S. F., Wittneben J. G., et al. (2012). Long-term stability of osseointegration in posterior maxilla using implant-supported fixed partial dentures: 5-year results of a prospective study. *International Journal of Oral and Maxillofacial Implants*, 27(6), 1231–1236.
49. O'Shea K., Bayraktar H. H., Silberschmidt V. V., et al. (2018). Experimental methods and numerical modeling techniques for characterization of dental implants: A review. *Materials Science and Engineering: Part C*, 93, 555–565.
50. Monopoli M. P., Aberg C., Salvati A., et al. (2012). Biomolecular coronas provide the biological identity of nanosized materials. *Nature Nanotechnology*, 7(12), 779–786.
51. Chen L., Zheng L., Wu B., et al. (2019). The effects of implant design on insertion torque, implant stability quotient, and bone-implant contact percentage: A systematic review and meta-analysis. *Journal of Prosthodontic Research*, 63(4), 361–371.
52. Chrcanovic B. R., Albrektsson T., Wennerberg A. (2018). Dental implants inserted in fresh extraction sockets versus healed sites: A systematic review and meta-analysis. *Journal of Dentistry*, 68, 1–7.
53. Albrektsson T., Wennerberg A. (2004). Influences of surface topography on implant integration: A literature review. *Journal of Periodontology*, 75(8), 148–157.
54. Fischer D. J., Nelson K., Switzer B., et al. (2016). Prevalence of metal hypersensitivity in patients with failed dental implants. *Clinical Oral Implants Research*, 27(9), 1116–1122.
55. Chen S. T., Buser D. (2014). Esthetic outcomes following immediate and early implant placement in the anterior maxilla—A systematic review. *The International Journal of Oral and Maxillofacial Implants*, 29(Supplement), 186–215.
56. Silva J. C., Senna P. M., Batista V. E., et al. (2022). Photobiomodulation therapy as an adjunctive treatment for peri-implant mucositis: A case series. *Journal of Clinical Periodontology*, 49(2), 235–242.
57. Schierano G., Bignardi C., Malacarne P. F., et al. (2021). Zirconia implant fractures: A systematic review and meta-analysis. *Journal of Dentistry*, 113, 103757.
58. Bumgardner J. D., Wally Z. J., Ong J. L., et al. (2017). Surface modification of implants for improved osseointegration. *Journal of Orthopaedic Research*, 35(4), 691–703.
59. Webster T. J. (2006). Nanophase ceramics for improved drug delivery and tissue engineering. *Journal of Nanoscience and Nanotechnology*, 6(7), 1961–1983.
60. Liskova J., Lassandro F., Fino P., et al. (2020). Mechanical behavior of dental implants and their interfaces: A review of experimental methods and finite element analysis. *Journal of the Mechanical Behavior of Biomedical Materials*, 110, 103876.
61. Eskow A. J., Mehta A., Carroll M. J., et al. (2015). The impact of implant surgical technique on early failure rates: A retrospective analysis of one provider's experience. *The International Journal of Oral and Maxillofacial Implants*, 30(3), 600–607.
62. Javed F., Al-Rasheed A., Al-Hezaimi K., et al. (2013). Effect of cigarette smoking on the outcome of dental implant: A systematic review and meta-analysis. *Journal of Periodontology*, 84(10), 1436–1446.
63. Mendonça G., Mendonça D. B., Aragão F. J., et al. (2013). Advancing dental implant surface technology - from micron- to nanotopography. *Biomaterials*, 34(28), 6929–6945.
64. Bandyopadhyay A., Bernard S., Xie J., et al. (2006). Influence of porosity on mechanical properties and in vivo response of Ti6Al4V implants. *Acta Biomaterialia*, 2(6), 677–690.
65. Geng W., Yan X., Zhang W., et al. (2019). Advances in 3D printing of porous metal implants. *Chinese Journal of Traumatology*, 22(6), 366–373.

66. Dohan Ehrenfest D. M., Del Corso M., Kang B. S., et al. (2010). The impact of the centrifuge characteristics and centrifugation protocols on the cells, growth factors, and fibrin architecture of a leukocyte- and platelet-rich fibrin (L-PRF) clot and membrane. *Platelets*, 21(4), 245–255.

67. Mangano F. G., Zecca P. A., Pozzi-Taubert S., et al. (2017). A new approach to full-mouth rehabilitation with dynamic navigation technology and intraoral scanners: A case series. *Journal of Esthetic and Restorative Dentistry*, 29(1), 43–52.

68. Esposito M., Hirsch J. M., Lekholm U., et al. (2000). Biological factors contributing to failures of osseointegrated oral implants. (II). Etiopathogenesis. *European Journal of Oral Sciences*, 108(6), 721–764.

69. Fasbinder D. J. (2012). Digital dentistry: Innovation for restorative treatment. *Compendium of Continuing Education in Dentistry*, 33(4), 246–248, 250–254.

70. Friberg B., Jemt T., Lekholm U. (1991). Early failures in 4,641 consecutively placed Brånemark dental implants: A study from stage 1 surgery to the connection of completed prostheses. *International Journal of Oral and Maxillofacial Implants*, 6(2), 142–146.

12 Sustainable Biomanufacturing

*Azan Ali, Mohammad Usman,
Elammaran Jayamani, KokHeng Soon,
and Mohamad Kahar bin Ab Wahab*

12.1 INTRODUCTION

12.1.1 Manufacturing

Manufacturing, defined as transforming materials and information into goods to satisfy human needs, is one of the world's primary sources of wealth [1]. Manufacturing encompasses two main processes: (1) the large-scale production of finished goods through the conversion of raw materials, and (2) the creation of complex products by supplying basic goods to manufacturers who use them to produce items such as household appliances, aircraft, and automobiles. A large part of the world's economy is based on manufacturing. Although consumption patterns differ slightly from one global territory to the next due to local cultural, social, and economic factors, the average global consumer expenditure continues to rise as living standards improve. Current global trends indicate that manufacturing can still serve as a possible means for growth and advancement, even for the most economically challenged countries [2]. The field of manufacturing offers exclusive opportunities for leveraging economies of scale, engaging in technological innovation and education, benefiting from the transfer of knowledge to other industries, and generating employment opportunities for a diverse range of skill sets.

12.1.2 Sustainable Biomanufacturing

Sustainable biomanufacturing uses living organisms, or biological systems, to produce products in an environmentally responsible and economically viable manner. According to Ehrenfeld [3], sustainability as a notion implies facilitating the capacity for life to flourish on our planet over a long period, that is, for many generations to come. The goal is to create products that minimize waste, reduce the environmental impact, and promote sustainable resource use. Biomanufacturing has grown rapidly in recent years due to increased demand for environmentally friendly and sustainable products, as well as advances in biotechnology that have made it easier to manipulate living organisms to produce specific products.

DOI: 10.1201/9781003375098-12

The key principle of sustainable biomanufacturing is to create a closed-loop system that minimizes waste and pollution while maximizing the use of renewable resources. This is achieved using biological systems such as bacteria, yeast, algae, and fungi, which can be used to produce a range of products, from bioplastics to biofuels.

The importance of biomanufacturing lies in its ability to produce high-value products with high purity, specificity, and efficacy. Compared to traditional manufacturing processes, biomanufacturing offers many advantages, such as scalability, flexibility, and reduced environmental impact. Additionally, biomanufacturing can produce complex molecules, including proteins and antibodies, which are difficult or impossible to produce using chemical synthesis.

Sustainable biomanufacturing also has social and economic benefits. For example, it can create new jobs in the green economy and provide economic opportunities for farmers and other rural communities. Additionally, it can promote food security by using agricultural waste and other biomass resources to create products, such as biofuels and bioplastics.

One example of sustainable biomanufacturing is using microorganisms, such as bacteria or yeast, to produce biofuels, chemicals, and other products. These organisms can be engineered to convert renewable resources, such as plant matter or agricultural waste, into useful products, reducing the need for non-renewable energy and materials. Additionally, biomanufacturing processes often generate less waste than traditional manufacturing processes, which can be toxic and harmful to the environment.

Another example of sustainable biomanufacturing is using plants, such as algae or moss, to produce high-value compounds, such as pharmaceuticals or fragrances. These organisms can be grown using renewable resources, such as sunlight and carbon dioxide, and harvested and processed to extract the desired product. Biomanufacturing processes using plants also have the advantage of producing a wide range of products, making them a flexible and scalable solution for sustainable manufacturing.

In summary, sustainable biomanufacturing is a process that aims to create environmentally sustainable, socially responsible, and economically viable products using renewable resources and closed-loop systems. This process minimizes waste and pollution, creates new economic opportunities, and promotes food security. To achieve sustainable biomanufacturing, it is necessary to optimize the use of renewable resources, minimize waste and pollution, use genetically modified organisms (GMOs) to increase efficiency, consider the entire life cycle of the product, and consider social and economic factors.

12.2   EVOLUTION AND HISTORY OF MANUFACTURING

Each step of manufacturing from the creation of new ideas to the improvement of existing techniques and the introduction of new machinery is constantly subject to change. Manufacturing in the context of industrial production typically refers to the process by which raw materials are transformed into finished products for sale on the market. On the other hand, modern manufacturing is viewed as a holistic enterprise that encompasses everything from individual machines and production lines to supply chains and even entire countries [4]. Over the course of several decades, improvements in the manufactured products of technology and machine tools along with the strategies of organizations aimed at cutting costs, boosting quality and reliability, increasing productivity and profits, and fostering sustainability have all contributed to the development of modern manufacturing systems. Several manufacturing system paradigms have emerged in response to the ever-increasing product variety and the subsequent need to differentiate in order to remain competitive. The need to manage fluctuations in production volumes and product varieties drove the early evolution of manufacturing systems from dedicated to flexible and reconfigurable [5].

12.3 APPLICATIONS OF BIOMANUFACTURING

12.3.1 BIOFUELS

Biomanufacturing produces biofuels using microorganisms, such as bacteria or yeast, to convert renewable resources, such as plant matter or agricultural waste, into fuel. The microorganisms are typically engineered through genetic modification to produce specific enzymes that break down the renewable resources into simple sugars, which can then be fermented to produce biofuels such as ethanol or butanol. Carbohydrases and ligninases are the two types of pretreatment enzymes. Carbohydrases breaks down complex carbohydrates such as cellulose and hemicellulose found in feedstock materials into simpler fermentable sugars such as glucose and xylose. Ligninases, on the other hand, degrade lignin, a complex polymer that creates a barrier around cellulose, making enzymes difficult to reach and degrade it. As a result, ligninases play an important role in improving carbohydrate conversion efficiency by eliminating lignin barriers that may restrict cellulose accessibility. Recent studies have shown that using a combination of different pretreatment enzymes can significantly enhance the efficiency of biofuels production. For example, a study by Zhao et al. [6] found that a combination of cellulases, xylanases, and pectinases significantly improved the efficiency of biofuels production from corn stover. Similarly, another study by Ilić et al. [7] found that a combination of cellulases and ligninases improved the efficiency of biofuels production from lignocellulosic biomass.

One example of biomanufacturing for biofuel production is the use of algae, which are known for their ability to produce high amounts of lipids that can be converted into biofuels. Algae are grown in photobioreactors and can be harvested to extract lipids, which can then be converted into biofuels through a process of transesterification [8]. Biofuels, such as algae fuels, have up to 45% O_2 content and very low sulfur emissions, whereas fossil fuels have almost no O_2 content and very high sulfur emissions [9]. Biofuels are non-polluting, widely available, sustainable, and reliable fuels derived from renewable resources. Microalgae-based fuels appear to be ecologically benign and non-hazardous with great potential for lowering global CO_2 emissions.

Another example is the use of bacteria to produce biofuels from waste streams, such as wastewater or agricultural waste. Wastewater includes a high concentration of organic materials, such as fats, oils, and grease, which may be utilized as feed-stock for biofuel generation. Bacteria are utilized to break down organic materials in a process known as anaerobic digestion. The bacteria are engineered to produce specific enzymes that break down the waste into simple sugars, which can then be fermented to produce biofuels. This process not only produces biofuels but also helps to reduce waste and lower the environmental impact of waste disposal [10].

Methanogenic bacteria are a kind of bacterium that is often utilized in anaerobic digestion. These bacteria can degrade organic substances and produce methane, which may be used as a biofuel. Methanogenic bacteria are naturally present in animal digestive systems, but they may also be grown in the laboratory for use in anaerobic digestion.

Biomanufacturing has proven to be a valuable tool for producing biofuels in a sustainable and environmentally friendly manner. By using microorganisms to convert renewable resources into biofuels, biomanufacturing provides a scalable solution to meet the increasing demand for sustainable energy sources. However, the ability of microalgae to serve as a productive raw material for biofuel production needs continued efforts to overcome numerous hurdles such as cultivating algae, creating sugar-rich biomass, harvesting, drying, biomass pretreatment, and ensuring maximum yielding processes.

12.3.2 Biopharmaceutical Manufacturing

Biopharmaceuticals, which are derived from biological sources, play a crucial role in modern healthcare. The process of manufacturing biopharmaceuticals, including enzymes, vaccines, antibodies, growth factors, and hormones, is challenging and requires strict control over production conditions to ensure consistent quality, efficacy, safety, and yield. A noteworthy example of this is the antibacterial properties of marine algae, which have been reported to be effective against various bacterial and fungal strains, including *Escherichia coli*, *Salmonella typhoid*, *Staphylococcus aureus*, and *Enterococcus faecalis*, as reported by Alsenani et al. [11] due to the antibacterial properties of marine algae potential opportunities for developing new biopharmaceutical treatments, but it also highlights the need for careful monitoring and control during production to ensure product quality and safety. The potential of microalgae and cyanobacteria as rich sources of bioactive compounds with diverse properties has led to increased interest among manufacturers and researchers. These organisms are expected to become a valuable commercial source for producing high-value bioactive compounds. Microalgae and cyanobacteria are attracting growing interest among manufacturers and researchers due to their potential as rich sources of bioactive compounds with diverse properties. These organisms are expected to play an increasingly important role in the commercial production of high-value bioactive compounds. The abundance of microalgae and cyanobacteria, coupled with their ability to synthesize a wide range of compounds, make them an attractive source for biopharmaceutical production, as well as for other applications such as cosmetics and nutraceuticals.

Biopharmaceuticals are designed to enhance the quality of life and life expectancy of patients with severe disorders, typically youngsters. However, many established economic and scientific practices in the pharmaceutical industry, particularly the biopharmaceutical sector, have worked to degrade the environment and undermine public confidence in the chase for profit [12]. The aquatic environment has been negatively impacted by the increase in pharmaceutical consumption and consequent discharge into wastewater, as it disrupts biological processes and causes harm to aquatic species, as noted by Fabbri [13]. Additionally, the presence of pharmaceutical effluents has been associated with the emergence and propagation of microorganisms that are resistant to antimicrobial agents, according to Milmo [14]. Moreover, to achieve a complex obstacle, the latter cost is extremely high for the manufacturing process. Biopharmaceutical pricing has been identified as a universal problem with impacts on high-income and low-income countries alike [15].

To boost operational efficiency and minimize facility costs, the biopharmaceutical sector is investing in breakthrough technologies such as single-use technology (SUT) to adopt process intensification, continuous manufacturing (CM), and closed processing.

12.3.3 SINGLE-USE TECHNOLOGY

The first form of the biopharmaceutical processing system is single-use systems, which are meant to be used for the duration of the production process of a single batch of therapies and then destroyed. The increased usage of single-use technology is lowering the danger of product cross-contamination by reducing the requirement for cleaning between batches. Manufactured in a cleanroom environment, these systems are double-bagged and sterilized using gamma, ethylene oxide (EtO), or x-ray sterilization techniques, ensuring sterility for every batch produced. This process is not only highly efficient but also cost-effective. Single-use technologies are appealing due to additional benefits such as lower capital costs and the possibility of debottlenecking during facility changeover [16]. A study by MilliporeSigma discovered that SUT can cut process development durations by up to 50% while improving process yields by up to 50% [17]. Since the introduction of the first single-use bioreactor (SUB), SUT has revolutionized biomanufacturing in several ways. Continuous SUT innovation has resulted in fully closed processes combined with process intensification by perfusion or CM bioprocessing, resulting in higher production at lower cost than previous solutions. As a result, biomanufacturing facilities are simpler, less expensive, and faster to install, with better flexibility and agility than standard stainless steel (SS) equipment.

SUT is widely utilized in biomanufacturing, and its use is projected to increase. Markets estimate that the global market for SUT in biopharmaceutical manufacturing will reach $5.44 billion by 2023, increasing at a compound annual growth rate (CAGR) of 17.3% from 2018 to 2023 [18].

12.3.4 CONTINUOUS BIOMANUFACTURING (SUT)

In contrast to traditional batch manufacturing, which comprises different manufacturing phases, continuous manufacturing allows for the smooth and uninterrupted manufacture of a product. The application of continuous manufacturing in the bio-pharmaceutical industry has gained much interest because of its potential to increase efficiency, lower costs, and improve product quality.

In biopharmaceuticals, continuous manufacturing combines many unit processes, such as cell culture, purification, and formulation, into a single continuous process. This enables continuous material flow, real-time process monitoring and control, and the capacity to make real-time modifications to ensure product quality. Continuous production has the benefit of producing things more effectively and with less waste than traditional batch manufacturing. Continuous production eliminates the need for intermediate storage and shortens the time necessary for process stages, resulting in faster product turnover.

In general, continuous procedures increase productivity or the production rate considerably; however, the increase in productivity or rate is sometimes accompanied by a loss(es) in product titer, specific titer (product titer/cell density), conversion yield, or other relevant fermentation parameters [19]. The adoption of continuous biomanufacturing presents a range of hurdles, including the potential for contamination over extended periods of continuous operation; genetic instability or mutation during ongoing culture; the challenge of maintaining optimal conditions for both cell growth and product formation concurrently; and a lack of effective tools and approaches for ongoing biomanufacturing research and development. In order to overcome these challenges, it is necessary to develop new strain design and fermentation engineering methodologies that can decrease the risk of contamination, improve genetic stability, and maintain high product yield while ensuring optimal conditions for cell growth and product formation.

12.3.5 Closed-Process Biomanufacturing (SUT)

Closed processing involves operating a product and material flow channel in a closed system, which can greatly reduce or eliminate the risk of environmental contamination and ensure product quality. Both stainless steel and single-use manufacturing facilities can benefit from closed processing. However, the potential for single-use processes to be closed using tube welders and aseptic connections has attracted considerable attention. Closed processing offers significant quality advantages over open processes or connections in a clean-room environment, as it eliminates the possibility of contaminants from the production environment. In addition, closed systems may provide economic benefits by eliminating the need for clean room–based production and reducing costs associated with gowning, cleaning, monitoring, and heating, ventilation, and air-conditioning (HVAC), as highlighted by Hodge [20].

Using single-use bioreactors is an example of closed-process production in biomanufacturing. These systems are intended for cell and microbe culture in a closed environment, eliminating the need for cleaning and sterilizing between batches. SUBs also provide various advantages over standard stainless steel bioreactors, including cheaper capital costs, reduced danger of cross-contamination, and greater flexibility in production scale-up [21].

12.4 BIOMATERIALS

Biomaterials are naturally occurring or man-made materials such as polymeric polymers, metals, ceramic materials, and their compounds that are easily integrated into the human body and are highly biocompatible with living tissues [22]. Plant-based materials such as cellulose, chitin, and lignin, as well as animal-based materials such as silk and collagen, are among the numerous types of sustainable biomaterials being produced and studied. These materials have several potential uses, ranging from packaging to textiles, medical devices, and building materials.

Biomaterials can be effectively implanted as heart valves, dental implants, vascular grafts, lenses, intraocular ligaments, tendons, etc. Certain medical equipment, such as pacemakers, biosensors, artificial skin, and artificial hearts, can be effectively made utilizing biomaterials [23]. Biomaterials are utilized to enhance or replace damaged or deteriorating organs and tissues. Biomaterials, which are materials designed to interact with biological systems, exhibit unique physical, chemical, and biological properties. As a result, a wide range of materials such as biodegradable polymeric materials, ceramic materials, metals and alloys, biopolymers, composite materials, biocomposite materials strengthened with natural fibers, and other filler materials are commonly used in the field of biomaterials. These materials are selected based on their ability to be processed, their mechanical and chemical properties, their compatibility with biological systems, and their ability to degrade or biodegrade over time.

12.4.1 DIFFERENT TYPES OF BIOMATERIALS

12.4.1.1 Metals

Utilizing biodegradable metals instead of permanent metallic implants may result in considerably superior fracture-fixing methods for situations where full tissue regeneration is envisaged. Currently, magnesium (Mg), iron (Fe), and zinc (Zn) alloys are the best-explored biodegradable metals for orthopedic and cardiovascular applications. Mg is a biodegradable material that exhibits high specific strength and low elastic modulus, similar to bone. Its density is also low at 1.74 g/cm^3. These properties reduce the risk of stress shielding, making it a promising material for biomedical applications. Iron is a biomaterial that is used in biomedical applications due to its biocompatibility and mechanical properties. It is naturally found in the human body and plays a crucial role in various biological processes. Iron-based biomaterials, such as iron oxide nanoparticles, have been used for drug delivery, imaging, and magnetic hyperthermia therapy. However, the biodegradability of iron-based biomaterials can be a concern as they may degrade too quickly or cause inflammation. Therefore, further research is needed to fully understand the potential of iron as a biomaterial and to optimize its properties for specific biomedical applications.

12.4.1.2 Biopolymers

Polymer-based biomaterials have found wide applications in the fields of medicine, biotechnology, food, and cosmetics. They are used in the production of various medical devices and products, such as implants, wound dressings, sutures, catheters, and stents. Biopolymer-based biomaterials have also been used in ligament and tendon repairs, as well as in the manufacture of valves used in cardiac surgeries.

Polymeric biomaterials used in biomedical applications are generally classified as synthetic, natural, or a combination of both polymers. Natural polymers are derived from plant and animal sources, including silk, wool, DNA, RNA, cellulose, and proteins. These natural polymers have attracted interest due to their biocompatibility, biodegradability, and low toxicity. They have been used in various medical applications, such as tissue engineering and drug delivery. However, their properties need to be optimized for specific applications to improve their effectiveness and stability.

Cellulose, chitin, and chitosan are structurally related, renewable, low-cost biopolymers that are growing in importance as the demand for sustainable alternatives to synthetic polymers grows. At the molecular level, their functional groups (OH, CH_3CONH-, and $-NH_2$) interact significantly, resulting in the formation of rather homogeneous nanocomposite (NC). They are most easily produced by mixing biopolymer solutions, followed by product regeneration into various physical "shapes", such as fibers, films, and hydrogels.

12.4.1.3   Bioceramics

Bioceramics exhibit varying levels of biocompatibility depending on their composition. Some ceramic oxides are inert in the body, while the body eventually replaces other resorbable materials after aiding in the repair process. Bioceramics are used in medical devices as rigid materials for surgical implants. Some bioceramics are flexible, mimicking the body's own materials, while others are extremely durable metal oxides. They find common use in dental and bone implants, joint replacements, and surgical cermet. Bioceramic coatings are sometimes applied to medical devices to reduce wear and inflammatory responses.

12.4.1.4   Biocomposites

To reduce the reliance on fiberglass composites, researchers are exploring alternative materials such as biocomposites and biopolymers. These materials involve the combination of synthetic polymers reinforced with natural fibers such as cotton, flax, hemp, or fibers with a polymer matrix from renewable and non-renewable sources to form composite materials. Ideally, the matrices used in these materials should be made from green sources such as vegetable oils or starches. These natural composites, made from biofiber and petroleum-derived non-biodegradable plastics (polyethylene and polypropylene) or biodegradable plastics (polyethylene and polypropylene), are known as biocomposites (PLA, PHA). Biopolymers, also referred to as green composites, are made from plant-derived fibers (natural/biological fibers) and crops/biological-derived plastics (biopolymers/bioplastics) and are a broader classification of biomaterials. The use of these materials could have potential benefits for the community [24].

12.4.2   3D Printing

3D printing, also known as additive manufacturing, has revolutionized biomanufacturing by creating complex 3D structures with precise control over the composition and architecture of the final product. By employing 3D printing to generate customized living materials, microbes may be linked to material creation, which can sustain microbial viability and metabolic activity in biological fermentation processes. Current microbial suspension fermentation techniques are neither portable, reusable, or even suited for on-demand biochemical synthesis. Through 3D printing technology, sustainable and tailor-made living materials can be specially developed and created, which can retain microbial activity to create reproducibly valuable products. A commercial 3D printer, for example, has been customized to produce patterned alginate hydrogel containing embedded *Escherichia coli.*

In 3D printing, metal implant parts can be produced through a layer-by-layer printing process using computer-aided designs (CADs). Metal 3D printing involves the layer-by-layer printing of metallic powders, which are then fused or sintered together using a laser technique. The two most commonly used laser sintering techniques for metal 3D printing are selective laser melting (SLM) and direct metal laser sintering (DMLS). These techniques allow for the production of complex metal parts with high precision and accuracy. However, it is important to note that the use of these techniques requires specialized equipment and expertise.

The fabrication of scaffolds for tissue engineering is one of the most important uses of 3D printing in biomanufacturing. These scaffolds can be designed to mimic the structure and properties of natural tissues, providing a framework for the growth of cells and tissues. 3D bioprinting offers distinct technological benefits in that it allows for exact spatiotemporal positioning of cells, ECM, biomaterials, and growth factors within 3D bioengineered constructs/scaffolds, resulting in biomimetic tissue and organ analogs for various biomedical applications [25].

In research conducted by Adams et al. [26], a soft human kidney phantom, featuring the pyelocaliceal system and full anatomy, has been developed through a distinctive production process that involves the combination of polymer molding and 3D printing. This method is highly versatile as it allows for the replication of anatomical characteristics with sub-millimeter accuracy and facilitates the use of various materials, including biocompatible hydrogels.

12.5 MICROBIOMES FOR SUSTAINABLE BIOMANUFACTURING

Microbiomes are complex microbial ecosystems that play important roles in biomanufacturing processes. Microbiomes (also known as microbial communities) are abundant in nature and have been utilized for direct human advantages such as pollution removal and renewable energy production. Microbiomes have several advantages over monoculture systems, including enhanced robustness owing to functional degeneracy and the capacity to alter a wide range of substrates. They can influence product quality, productivity, and efficiency. Using microbiomes for sustainable biomanufacturing has emerged as a promising approach to reducing environmental impacts, improving yields, and enhancing product quality.

For biomanufacturing, microbiomes offer numerous distinct benefits over pure cultures, the most prominent of which is the division of labor. The division of labor in microbiomes happens when multiple species/strains collaborate in performing a metabolic pathway rather than a single species performing it exclusively. This cooperation speeds up the flow toward the final product by reducing the negative impact of intracellular resource constraints, such as competition for energy resources (such as Adenosine triphosphate [ATP]), cell space, protein expression machinery (e.g., ribosomes and RNA polymerase), and inhibitory metabolic intermediates [27].

Apart from the division of labor, functional degeneracy is another important feature that makes microbiomes more appealing for biomanufacturing than pure cultures. When many species execute duplicate fundamental activities or processes but have varying fitness levels regarding other physiological features (temperature and pH optima, biofilm formation capacity, resistance to harmful chemicals, etc.), functional degeneracy arises in microbiomes [28].

12.5.1 Microbiome Role in Different Sectors

12.5.1.1 Human Health

Microbiome biomanufacturing has the potential to transform the healthcare industry by enabling the creation of novel medicines, diagnostics, and prevention techniques. The microbiome is a colony of microorganisms that live in and on the human body and play a vital role in health maintenance and illness prevention.

The production of tailored probiotics is one possible use of microbiome biomanufacturing. Probiotics are living microorganisms that, when ingested in sufficient quantities, give health benefits. It may be possible to develop a tailored probiotic that addresses specific health issues by examining an individual's microbiome. In general, microbial composition differs across different anatomical areas, and it is highly individualized since the makeup of the microbiome varies among individuals. Although the precise definition of healthy microbiota has yet to be determined, studies have indicated that the use of probiotics, prebiotics, and synbiotics is advantageous by maintaining healthy body flora or by modifying the microbiome toward a healthy microbial ecology. Probiotics are live microorganisms that confer health benefits when consumed in adequate amounts. Biomanufacturing techniques can be used to develop probiotic strains that can improve gut health, reduce inflammation, and prevent or treat gastrointestinal disorders. For example, the probiotic strain *Lactobacillus rhamnosus* GG has been shown to reduce the incidence and severity of antibiotic-associated diarrhea in children [29].

On the other hand, prebiotics are non-digestible dietary fibers that promote the growth of beneficial bacteria in the gut. Biomanufacturing technologies can be used to produce prebiotic compounds that can selectively stimulate the growth of specific bacterial species, leading to a more diverse and healthier microbiome. For example, fructooligosaccharides (FOS) and galactooligosaccharides (GOS) are prebiotic compounds that have been shown to increase the abundance of Bifidobacteria in the gut [30].

Biomanufacturing technologies can also produce microbial-based therapeutics, which can directly modulate the microbiome and improve human health. For example,

Akkermansia muciniphila is a beneficial gut bacterium shown to reduce inflammation, improve metabolic health, and enhance immune function. Biomanufacturing techniques can be used to produce pure cultures of *A. muciniphila*, which can be administered as a therapeutic agent to treat various conditions [31].

12.5.1.2 Agriculture

The fields of biotechnology and microbiology collaborate to enhance crop production, crop quality, and the sustainability of existing systems, resulting in the generation of superior and more abundant agricultural products through the utilization of genetically modified organisms and transgenic crops. Microbial biotechnology has played an important part in sustainable agriculture, including biofertilizers, biopesticides, bioherbicides, and insecticides. Microbial biotechnology minimizes pesticide dependency in sustainable agriculture by managing biotic and abiotic stressors.

Plant microbiomes are important bioresources in agriculture, as they contain beneficial microbes that can enhance plant growth and improve plant nutrition uptake through various mechanisms. For example, some microbes in the plant microbiome can solubilize essential nutrients such as phosphorus, potassium, and zinc. In contrast, others can fix atmospheric nitrogen, an essential nutrient for plant growth [32].

In addition to these mechanisms, plant microbiomes can also play a role in biofortification, which involves increasing the nutrient content of crops through microbial activity. For example, some microbes in the plant microbiome can produce siderophores, molecules that bind to and transport iron, an essential nutrient for plant growth, and can improve iron uptake in plants.

12.5.2 CHALLENGES

While the potential of microbiomes for sustainable biomanufacturing is vast, several challenges must be addressed. Some challenges are as follows:

Contamination: Microbiomes can be prone to contamination, and it can be challenging to maintain the stability of a microbial consortium over time. Contamination can result in decreased productivity and the production of unwanted compounds [33].

Scale-up: Microbiomes can be challenging to scale up for commercial production. The conditions that optimize the performance of a microbial consortium at a laboratory scale may not be applicable at larger scales, and it can be challenging to maintain the stability and composition of the consortium over time.

Difficulty in predicting performance: It can be challenging to predict the performance of a microbial consortium, as it depends on the interactions between different microorganisms and environmental factors. This can make it difficult to optimize biomanufacturing processes.

Lack of suitable hosts: Some microbiomes may require specific hosts for their function, and it can be challenging to identify and optimize suitable hosts for biomanufacturing. This can limit the potential applications of microbiomes in biomanufacturing.

Variable performance: The performance of a microbial consortium can be variable over time, and it can be challenging to maintain consistency in the production of a specific compound. This can result in batch-to-batch variability and make it difficult to meet quality control (QC) standards.

12.6 CURRENT CHALLENGES AND POTENTIAL SOLUTIONS OF SUSTAINABLE BIOMANUFACTURING

12.6.1 HIGH COST

Challenge description: The high cost of manufacturing cell therapy and tissue-engineered organs under current good manufacturing practice (cGMP) conditions for clinical use makes it difficult to ensure widespread availability of these treatments [34].

Potential solution: Streamlining cell manufacturing through the creation of a universal, specified medium. A standardized bio-ink for 3D printing.

12.6.2 INSUFFICIENT AUTOMATION

Challenge description: Many biomanufacturing processes are carried out manually and are not entirely automated.

Potential solution: Develop procedures for biomanufacturing that are completely automated and can be employed with multiple cell types for cell treatments or tissue-engineered products.

12.6.3 UNDERDEVELOPED QC

Challenge description: There is a particular requirement for non-destructive in-line sensing quality control.

Potential solution: Develop non-destructive in-line technology that can be included in automated procedures in order to guarantee the quality attributes of the biomanufactured clinical product.

12.6.4 LACK OF SUFFICIENT STANDARDIZATION

Challenge description: The lack of standards for biomanufacturing accessory materials is a particular deficiency.

Potential solution: Consistent and high-quality auxiliary materials used in biomanufacturing require the establishment of standard criteria and supporting documentation.

12.6.5 LACK OF MODULAR SYSTEMS

Challenge description: There are no readily available, adaptable modular solutions for biomanufacturing at this time.

Potential solution: Create universally applicable modular bioreactors for the proliferation and maturation of bioengineered cells, tissues, and organs.

12.7 BLOCKCHAIN TECHNOLOGY IN BIOMANUFACTURING

12.7.1 Conception and Use of Blockchain in Biomanufacturing

The technology is intended to ease the internal movement of funds between parties without the need for financial institutions or middlemen. The primary aim of utilizing blockchain technology is to ensure the authenticity and safety of data and financial transactions, ensuring that all parties involved have access to complete and accurate information [35]. Biomanufacturing is a large field that encompasses a variety of low- to high-value products based on the size of the market. A bioproduct's medicinal potential is largely contingent on its level of purity. Consequently, both the production and purifying processes are crucial to obtaining the desired result [36,37]. The biopharmaceutical manufacturing process involves a range of techniques, including the fermentation and cell culture of both mammalian and plant cells, as well as recovery and purification steps. These purification steps involve a variety of critical chromatography processes, such as gas chromatography and high-performance liquid chromatography, as well as gel exclusion and affinity chromatography. In addition to chromatography, other advanced purification processes, such as centrifugation and ultrafiltration, are also employed. These processes are essential to ensure the quality, safety, and efficacy of the final product. While there is ongoing research to develop more efficient and cost-effective manufacturing processes, the current techniques have been refined over many years of experience and are considered standard practice in the biopharmaceutical industry [38]. The diverse architecture of bioreactor designs is also crucial for the efficient execution of biomanufacturing. Consequently, taking note of the vital process parameters throughout production is a crucial step for limiting the hazards associated with process purification operations. The complex of biomolecules and posttranslational alterations in the host constitute one of the primary obstacles to preserving the stability and usefulness of biological products. The process of biomanufacturing involves several crucial steps, including posttranslational modifications that ensure the bioproducts can withstand harsh environmental conditions and remain active. Given the complexity of biomanufacturing, an automated, transparent, immutable, and digitalized infrastructure is required to monitor and record various inputs and outputs of the process. Such an infrastructure can also aid in the dissemination of the final product to the consumer forum. Transparency and immutability ensure the traceability and auditability of the manufacturing process, which ultimately improves the quality and safety of the final product. Additionally, a digitalized infrastructure enables efficient tracking and monitoring of the product's distribution, ensuring it reaches the intended audience and is used safely and effectively. Overall, the implementation of an automated, transparent, immutable, and digitalized infrastructure can improve the efficiency, consistency, and reliability of biomanufacturing processes, ultimately leading to high-quality and effective bioproducts. As a result of examining these criteria, blockchain technology is ideally suited for this application since it pertains to many aspects to benefit the sustainability of biomanufacturing [39].

12.7.2   SIGNIFICANT CHARACTERISTICS OF BLOCKCHAIN
TECHNOLOGY IN THE ARENA OF BIOMANUFACTURING

12.7.2.1   Autonomy

Description of the function: This implies that any node of the blockchain may safely access, transfer, store, and update data.

Benefits: Sharing data with the public sector can also yield significant benefits while maintaining a high level of security and privacy.

12.7.2.2   Decentralized

Description of the function: Through blockchain technology, data can be viewed, monitored, stored, and updated across multiple machines, ensuring its integrity and accessibility.

Benefits: By utilizing blockchain technology, a manufacturer can access data and oversee any deviations from the specified protocols during the manufacturing process. Additionally, the supply chain can be more easily tracked and monitored, enhancing transparency and efficiency.

12.7.2.3   Transparent

Description of the function: This signifies that the data is visible to the reader, preventing the manipulation of data.

Benefits: Consumers and producers create a connection characterized by greater transparency and trust. A quality check is also considered feasible at each stage [40].

12.7.3   BLOCKCHAIN TECHNOLOGY IN THE DEVELOPMENT OF COVID-19 VACCINE

The disaster of the COVID-19 epidemic, which has had severe effects on economic, social, and public health on a global scale, has ravaged the whole planet. The global advancement of vaccines has been achieved through the cooperative effort of the research institutions and private biotech companies. As a result, a new level of sustainable manufacturing and delivery capacity is required [41]. It is unquestionably clear that production and supply chain management play an equal role. As a result, substantial progress has been made toward a decentralized, blockchain technology distribution that has the capability of providing safe and secure encrypted data without any alteration to the actual information, thereby facilitating the growth of trust in the vaccine supply chain without relying on a third party, which includes the chance of risk or false information linked with the development and distribution of vaccine. Extensive study of the processes of viral replication, attachment, and binding is required for vaccine development. However, patient records, signs, and symptoms reveal more about the infection's course. Thus, a number of experimental research initiatives utilizing blockchain technology to collect and store data on a larger public forum have been developed [40]. The blockchain-based digitalized VIRI platform can resolve two crucial problems, namely data privacy and authenticity. In reality, the widespread use of blockchain technology for identifying virus carriers in different countries has contributed to the rapid detection and containment of viruses.

The platform securely stores user data and is coupled with artificial intelligence (AI) and machine learning (ML) capabilities, enabling accurate predictions of global outbreaks such as COVID-19 [42].

12.7.4 FUTURE PROSPECTS OF BIOTECHNOLOGY IN BIOMANUFACTURING

This technology's potential in the biomanufacturing industry still needs to be evaluated, despite the fact that the growing applications of blockchain will unquestionably affect all facets of life in the future and lead to a shift for the better. Although blockchain technology has been integrated into several major corporations, it is currently in a phase of disillusionment and facing certain challenges. However, it is expected that in the near future, there will be various interoperable blockchain platforms that are easily accessible, scalable, and user-friendly. While features such as immutability, transparency, and peer-to-peer networking in both private and public modes already exist, there are still questions regarding the implementation costs and universal adoption of these features. Despite these challenges, the potential benefits of blockchain technology are significant and ongoing developments in the field will likely lead to more widespread adoption and usage in various industries. It is expected that blockchain technology will soon enable two-way communication in the biomanufacturing industry. An alternative approach is to combine artificial intelligence with the technology of blockchain to enhance bioprocess design, its sustainable features, and economic aspects. To further boost the technology's credibility and trustworthiness in the biomanufacturing industry, cybersecurity for the protection of proprietary research, development data, and sensitive data regarding different pathogenic microorganisms and their processes needs to be strengthened for the future to improve the reliability of biomanufacturing technology in the industry [40].

12.7.5 ENVIRONMENTAL DESIGN OF SUSTAINABLE BIOMANUFACTURING

When contemplating the reduction of an activity's environmental design impact, it is essential to define and quantify each category of environmental stress caused. The environmental impacts and benefits of biomanufacturing are often assessed based on a single outcome or a limited time frame. The adoption of a life cycle assessment (LCA) methodology provides a holistic assessment of the environmental impacts and benefits linked to materials, products, and manufacturing processes, covering a diverse range of indicators throughout the entire life cycle of a given product. This includes various stages such as raw material extraction, refining, fabrication of manufacturing components, their use in biomanufacturing, and final end-of-life treatment, with a broad range of indicators assessed [43]. To reduce the environmental impact of a product or service, efforts should be made at every stage of the entire value chain, from discovery to manufacture of active ingredients to packaging, distribution, and disposal. This approach involves minimizing both direct and indirect emissions through various measures. Direct measures can include the electrification of processes and the selection of refrigerants that have lower global warming potential. For example, companies can use renewable energy sources such as solar and wind power to generate electricity for their manufacturing processes. Additionally, switching to refrigerants that have lower global warming potential, such as hydro-fluoroolefins (HFOs), can significantly reduce the environmental impact of refrig-eration and air-conditioning systems. Indirect emissions can be reduced through measures such as building and process optimization. This can involve improving energy efficiency in buildings by upgrading insulation, installing energy-efficient lighting and HVAC systems, and implementing smart building controls. In manufac-turing processes, optimizing processes to reduce waste and improve efficiency can also help reduce indirect emissions. Reducing the environmental impact of products and services requires a comprehensive approach that involves all stakeholders in the value chain. By taking steps to minimize emissions at every stage of the value chain, companies can make incremental progress toward decarbonization. The Bio-Process Systems Alliance (BPSA) is continuing with a series of publications analyzing the present condition and the relative environmental impact of various biomanufacturing materials [44,45]. They describe the 7Rs of a greener bioprocess, which are a set of guidelines for making decisions on how to create, manage resources, processes, and their eventual disposal. Reusing resources, for instance, reduces the overall impact on the environment by spreading the cost across multiple applications.

Assessing the existing situation and striving to improve the long-term viability of biomanufacturing activities require careful consideration of numerous factors. Safe and environmentally sound management of healthcare waste are essential to prevent negative health and environment implications, such as the unintentional release of chemicals or biological risks. For instance, cremation reduces contamination on land but releases carbon into the atmosphere. However, if one of the novel carbon capture methods becomes feasible, the scenario could change. In evaluating the costs and burdens associated with various available solutions, factors such as plant geographi-cal setting coefficients, distance to recycling plants, and the environmental impact of local power generation are taken into account. Additionally, national and regional laws, corporate objectives, customer expectations, and societal requirements are considered.

In order to achieve sustainability in production facilities, the HVAC design is a critical aspect that must be considered. While the current good manufacturing practice HVAC classifications and typical air change rate bands are used to deter-mine air change rates and environmental controls, during the initial stages of design, constraints such as process and operator occupancy and activity may not always be well defined, yet they can significantly influence these variables. To optimize HVAC systems for these factors, two approaches can be taken. One is to conduct thorough research and analysis of the specific operational needs and requirements of the facil-ity in question, and to use this information to create a customized HVAC design that accounts for all relevant variables. The other is to implement adaptive HVAC control systems that can dynamically adjust to changes in occupancy and activity levels to maintain optimal environmental conditions while minimizing energy consumption.

Taking more time on the design, gathering more precise data on the process/facility design to back up the equipment rightsizing, posing the proper questions, and determining the best air-handling rates.

Two rounds of commissioning, the first for the initial assessment of the building and the second to collect performance data for the purposes of optimizing the building.

The geographical location of a facility or industry is also an important factor in determining the sustainability of the design of biomanufacturing products. The environmental impact of an operation can be heavily influenced by various factors, including the proximity to energy sources, the use of greener energy grids, year-round weather conditions, the shipping of materials such as SU systems, final product logistics, and the availability of qualified staff. Facility design plays a crucial role in minimizing the environmental impact of an operation, and there is a trend toward a full, digital analysis of an operation's environmental burdens. In facility design, several operational parameters can be optimized to reduce the environmental impact of an operation. These parameters can include fabric U-values, electric loads, lighting technologies, and heating, cooling, and refrigerant chemistries. The integration of renewables, such as solar panels and wind turbines, can also significantly reduce the carbon footprint of an operation. Geographical building settings, including site weather data and service sources, can also be optimized to reduce the environmental impact. For example, a building's orientation can be designed to maximize natural light and reduce the need for artificial lighting, and the building's location can be chosen to reduce transportation emissions associated with material shipping and product logistics [46].

12.8   SUSTAINABLE BIOMANUFACTURING IN SOLAR ENERGY AND PHOTOSYNTHESIS

12.8.1   Development of Solar Energy in Biomanufacturing

The production capacity of renewable energy generated through solar panels is rapidly increasing, while solar thermal collectors are also contributing to the supply of heat [47]. Synthetic fuel systems that use solar energy to transform water and carbon dioxide into methanol, green liquid hydrogen, and green ammonia are at the forefront of renewable fuels research and development. Yet, a fourth route for solar energy usage is emerging: direct-light biomanufacturer. Photosynthetic organisms use light energy to split carbon dioxide, water, and minerals into oxygen and other proteins that make up biomass. The fourth sustainable biomanufacturing method has the potential to produce high-value compounds in large quantities while promoting sustainable production, in addition to biomass production. This is a crucial factor in supporting both aerobic life on earth and the global economy, as it ensures that resources are used efficiently and effectively for the long-term benefit of both the planet and its inhabitants [48].

Solar radiation continues to be an essential source of renewable energy for the future, despite its low photosynthetic energy-conversion efficiency of 1%–5% [49]. Additionally, ongoing advancements in solar technology are improving its efficiency and decreasing its cost, making it a more feasible and accessible option for many communities worldwide. By harnessing the power of solar radiation, we can decrease our reliance on non-renewable energy sources, mitigate the impacts of climate change, and promote a more sustainable future. Light energy is transformed into a wide variety of useful and structurally complex biological compounds, and plants and microalgae play a pivotal role in mediating the process [49].

12.8.2 SUSTAINABILITY OF PHOTOSYNTHETIC ORGANISMS

Photosynthetic organisms exhibit remarkable diversity, spanning from unicellular cyanobacteria and green algae to towering flowering trees. In fact, there are over 400,000 known species of green plants alone, highlighting the vast array of photosynthetic life that exists on our planet. This diversity is critical to maintaining the health of ecosystems, as different photosynthetic organisms are adapted to thrive in various environments and provide essential ecosystem services, such as oxygen production and carbon sequestration [50]. This variety has gained and continues to acquire profound adaptions to fluctuating settings over time [51]. To conserve this planet, it is mandatory to preserve photosynthesis-based ecosystems in nature and observe inspiration from them to redesign human-made production systems to run on less-polluting fuels. By creating resource-efficient and low-carbon circular bioeconomies, the earth can be transformed into a green plant where the effects of human life are mitigated [52–54].

Photosynthesis enables plants and algae to be robust, environmentally friendly, and global providers of food and materials. Vitamins, medications, stimulants, tastes, sauces, colors, and fine chemicals are just a few of the many in planta functions of bioactive natural compounds in plant algae. Photosynthetic organisms are crucial to the manufacturing process of bio-based civilizations that rely on renewable resources for their unique set of abilities. Hence, the engineering of photosynthesis may contribute to boost the revolutionary potential of photosynthetic output in the future [55].

12.8.3 STEM CELL BIOMANUFACTURING

Stem cells are becoming increasingly popular due to their ability to self-renew and differentiate into any desired cell type. Pluripotent, multipotent, and unipotent stem cells have gained significant attention for various applications, such as drug screening, disease modeling, and regenerative medicine. However, the current scope of stem cell–related biomanufacturing is mostly restricted to the production or creation of tissue engineering devices. While most of the existing research on biomanufacturing and stem cells is concerned with the design of bioreactors or the quality control of finished goods, the term "biofabrication" was coined by individuals interested in the industrial-scale manufacture of stem cell–linked medicinal products [56]. Biofabrication is a rapidly growing field of interest in tissue engineering and regenerative medicine. The use of automated methods, such as additive manufacturing technologies, to produce 3D biomaterial-based cell culture systems is a crucial aspect of biofabrication [57]. Biofabrication is an emerging field in tissue engineering and regenerative medicine that involves the use of automated techniques, including additive manufacturing technologies, to create 3D cell culture systems based on biomaterials. The primary focus of biofabrication is to produce hybrids of stem cells and biomaterials, such as scaffolds, microcarriers, and microgels. By utilizing various cell types and biomaterials, biofabrication emphasizes the fabrication methods to achieve a desired spatial organization of constructs that can provide an appropriate environment for stem cells to differentiate and mature after implantation. The ultimate goal of biofabrication in tissue engineering and regenerative medicine is to create functional and viable tissue substitutes that can repair or replace damaged or diseased tissues and organs.

The potential therapeutic uses of stem cells are exceptional. Yet, a large enough supply of stem cells is necessary to fully exploit their potential, making the question of how to mass-produce them a central one in the field of stem cell engineering. As mentioned above, bioreactors are commonly used in the production of stem cell products; however, due to the intricate nature of stem cell culture, it is difficult to directly translate 2D culture conditions into 3D bioreactor production. The interplay of biochemical variables, cell–cell interactions, and cell–matrix interactions in a culture dish determines the pluripotency and differentiation of stem cells [58] analyzed and evaluated the influence of hydrodynamic conditions on stem cell aggressions, metabolism, and phenotypic in suspension culture. Significantly, they viewed cell aggregates as a promising technique for large-scale synthesis. The emergence of fabrication technologies has made it possible to biomanufacture cell-based scaffolds with high precision and on a large scale through the use of computer-controlled bioprinting or bio assembly [59].

12.9 FUTURE OF SUSTAINABLE BIOMANUFACTURING

Biomanufacturing is certain to play a non-negligible but constrained role in absorbing and sequestering carbon from the atmosphere, given the market sizes and target compounds. The future of the medical industry is being shaped by various trends, including precision machining and control in micro-/nano-domains, advanced biomaterials, integration of medical subsystems, and personalized applications [60]. Rising expectations for improved quality of life have driven these trends. The market for these technologies is rapidly growing, and their application to medical industries is expected to bring about major breakthroughs, as seen in the machining industries.

In order to meet the needs of personalized medical applications, high-mix, low-volume production will be necessary. This can benefit from the lessons learned during previous paradigm shifts in other industrial sectors. In order to facilitate the entry of machining industry researchers and companies into the medical field, collaborations between academia, industry, and medicine must be encouraged. Biomedical engineers must possess a strong understanding of material and process mechanics and have access to advanced machine tools.

To foster innovation in the medical industry, attracting skilled individuals and promoting multidisciplinary collaborations are crucial, despite the current regulatory and cost requirements. Biomanufacturing provides an ideal platform for integrating advancements in precision engineering, nanotechnology, biotechnology, information technology, and cognitive sciences. Through these converging innovations, biomanufacturing can continue to push the limits of what is achievable in the medical field.

In order to encourage these converging innovations, it is important to foster an environment that attracts and retains talented individuals. This can be accomplished by offering competitive salaries, providing opportunities for professional development, and creating a culture of innovation and collaboration. In addition, promoting multidisciplinary collaborations can bring together experts from various fields to work toward a common goal. Furthermore, regulatory and cost requirements can be reduced through strategic partnerships with government agencies and private companies. By working together, these entities can develop new technologies and solutions that address both regulatory and cost challenges, ultimately leading to more efficient and effective biomanufacturing processes.

12.10   CONCLUSION

Sustainable biomanufacturing represents a promising approach for addressing some of the most pressing challenges facing modern society, including climate change, resource depletion, and public health concerns. By harnessing the power of biological systems and leveraging cutting-edge technology, it is possible to develop a more sustainable and efficient manufacturing model that delivers high-quality products while minimizing the environmental impact.

Despite the many potential benefits of sustainable biomanufacturing, however, there are also significant challenges and obstacles that must be addressed to fully realize its potential. These include regulatory barriers, technical limitations, and economic constraints, among others.

Nonetheless, there are many reasons to be optimistic about the future of sustainable biomanufacturing. Advances in genetic engineering, fermentation technology, and other key areas are providing researchers and manufacturers with powerful new tools and techniques for creating innovative bioproducts and improving existing ones. And as the global community becomes increasingly aware of the urgent need to transition to more sustainable and equitable modes of production, demand for biomanufacturing solutions is likely to continue growing. Overall, sustainable biomanufacturing is an exciting and rapidly evolving field with the potential to transform the way we produce and consume goods. By building on the principles of green chemistry, the circular economy, and biotechnology, we can create a more sustainable, equitable, and resilient future for all.

REFERENCES

1. Chryssolouris G, Papakostas N, Mavrikios D. A perspective on manufacturing strategy: Produce more with less. *CIRP J Manuf Sci Technol.* 2008;1(1):45–52.
2. Naudé W, Szirmai A. The importance of manufacturing in economic development: Past, present and future perspectives. MERIT Working Papers. 2012;2012–041.
3. Ehrenfeld J. *Sustainability by Design.* New Haven: Yale University Press; 2008.

4. Behzad E, Behdad S, Wang B. The evolution and future of manufacturing: A review. *J Manuf Syst.* 2016;39:79–100.

5. Elmaraghy HA, Laszlo M, Schuh G, Elmaraghy WH. Evolution and future of manufacturing systems. *CIRP Ann.* 2021;70(2):635–658.

6. Zhao Y, Damgaard A, Liu S, Chang H, Christensen TH. Bioethanol from corn stover – Integrated environmental impacts of alternative biotechnologies. *Resour Conserv Recycl.* 2020;155:104652.

7. Ilić N, Milić M, Beluhan S, Dimitrijević-Branković S. Cellulases: From lignocellulosic biomass to improved production. *Energies.* 2023;16(8):3598.

8. Borowitzka MA. *Microalgae: Biotechnology and Microbiology.* Cambridge: Cambridge University Press; 2013.

9. Hossain FM, Rainey TJ, Ristovski Z, Brown RJ. Performance and exhaust emissions of diesel engines using microalgae FAME and the prospects for microalgae HTL biocrude. *Renew Sustain Energy Rev.* 2018;82:4269–4278.

10. Biswas T, Bhushan S, Prajapati SK, Ray Chaudhuri S. An eco-friendly strategy for dairy wastewater remediation with high lipid microalgae-bacterial biomass production. *J Environ Manage.* 2021;286:112196.

11. Alsenani F, Tupally KR, Chua ET, Eltanahy E, Alsufyani H, Parekh HS, Schenk PM. Evaluation of microalgae and cyanobacteria as potential sources of antimicrobial compounds. *Saudi Pharm J.* 2020;28(12):1834–1841.

12. Lalor F, Fitzpatrick J, Sage C, Byrne E. Sustainability in the biopharmaceutical industry: Seeking a holistic perspective. *Biotechnol Adv.* 2019;37(5):698–707.

13. Fabbri E. Pharmaceuticals in the environment: Expected and unexpected effects on aquatic fauna. *Ann N Y Acad Sci.* 2014;1340(1):20–28.

14. Milmo S. Pharmaceutical environmental pollution and antimicrobial resistance. *Pharm Technol.* 2017;41(10):1–2.

15. Hurst DJ. Restoring a reputation: Invoking the UNESCO Universal Declaration on Bioethics and Human Rights to bear on pharmaceutical pricing. *Med Health Care Philos.* 2016;20(1):105–117.

16. Shukla AA, Gottschalk U. Single-use disposable technologies for biopharmaceutical manufacturing. *Trends Biotechnol.* 2013;31(3):147–154.

17. Press Release. MilliporeSigma Opens New Life Science Center in Massachusetts for Scientific Collaboration [Internet]. *Merckmillipore.com.* 2017 [cited 2023 Jul 2]. Available from: https://www.merckmillipore.com/INTL/en/20171010_231452?ReferrerURL=https%3A%2F%2Fwww.google.com%2F.

18. Single Use Bioprocessing Market Revenue Forecast. *MarketsandMarkets.* 2021 [cited 2023 Jul 2]. Available from: https://www.marketsandmarkets.com/Market-Reports/single-use-bioprocessing-market-231651297.html.

19. Xie D. Continuous biomanufacturing with microbes — Upstream progresses and challenges. *Curr Opin Biotechnol.* 2022;78:102793.

20. Hodge G. Points to consider in manufacturing operations. In *Biopharmaceutical Processing.* Elsevier; 2018, 987–998.

21. Lütke-Eversloh T, Rogge P. Biopharmaceutical manufacturing in single-use bioreactors current status and challenges from a CDMO perspective. *Pharm Ind.* 2018;80(2):281–284.

22. Biswal T, BadJena SK, Pradhan D. Sustainable biomaterials and their applications: A short review. *Mater Today Proc.* 2020;30:274–282.

23. Kulinets I. Biomaterials and their applications in medicine. *Regul Aff Biomater Med Devices.* 2015;1–10.

24. John MJ, Thomas S. Biofibres and biocomposites. *Carbohydr Polym.* 2008;71(3):343–364.

25. Li J, Wu C, Chu PK, Gelinsky M. 3D printing of hydrogels: Rational design strategies and emerging biomedical applications. *Mater Sci Eng R Rep.* 2020;140:100543.
26. Adams F, Qiu T, Mark A, Fritz B, Kramer L, Schlager D, Wetterauer U, Miernik A, Fischer P. Soft 3D-printed phantom of the human kidney with collecting system. *Ann Biomed Eng.* 2016;45(4):963–972.
27. Johnson DR, Goldschmidt F, Lilja EE, Ackermann M. Metabolic specialization and the assembly of microbial communities. *ISME J.* 2012;6(11):1985–1991.
28. Louca S, Polz MF, Mazel F, Albright MBN, Huber JA, O'Connor MI, et al. Function and functional redundancy in microbial systems. *Nat Ecol Evol.* 2018;2(6):936–943.
29. Szajewska H, Kolodziej M. Systematic review with meta-analysis: Lactobacillus rhamnosus GG in the prevention of antibiotic-associated diarrhoea in children and adults. 2015 Sep 13;42(10):1149–1157.
30. Davis LMG, Martínez I, Walter J, Hutkins R. A dose dependent impact of prebiotic galactooligosaccharides on the intestinal microbiota of healthy adults. *Int J Food Microbiol.* 2010;144(2):285–292.
31. Zhang G, Heng H. Human microbiome and environmental disease. *Environ Dis.* 2017;2(1):5.
32. Wu Z, Bañuelos GS, Lin Z-Q, Liu Y, Yuan L, Yin X, Li M. Biofortification and phytoremediation of selenium in China. *Front Plant Sci.* 2015;6:136.
33. García-Fraile P, Menéndez E, Rivas R. Role of bacterial biofertilizers in agriculture and forestry. *AIMS Bioeng.* 2015;2(3):183–205.
34. Hunsberger JG, Hickerson DHM. Biomanufacturing for regenerative medicine - Chapter 78. In *Principles of Tissue Engineering (Fifth Edition).* Academic Press; 2020, 1469–1480.
35. Ikeda K. Chapter seven - security and privacy of blockchain and quantum computation. *Adv Comput.* 2018;111:199–228.
36. Ahmad S, Kumar A, Hafeez A. Importance of data integrity & its regulation in pharmaceutical industry. *J Pharm Innov.* 2019;8(1):306–313.
37. Zhang YHP, Sun J, Ma Y. Biomanufacturing: History and perspective. *J Ind Microbiol Biotechnol.* 2017;44(4–5):773–784.
38. Aijaz A, Li M, Smith D, Khong D, LeBlon C, Fenton OS, Olabisi RM, Libutti S, Tischfield J, Maus MV, Deans R, Barcia RN, Anderson DG, Ritz J, Preti R, Parekkadan B. Biomanufacturing for clinically advanced cell therapies. *Nat Biomed Eng.* 2018;2(6):362–376.
39. Zobel-Roos S, Schmidt A, Uhlenbrock L, Ditz R, Köster D, Strube J. Digital twins in biomanufacturing. *Adv Biochem Eng Biotechnol.* 2020;176:181–262.
40. Pandey M, Singhal B. Blockchain technology in biomanufacturing. In *Blockchain Technology for Emerging Applications.* Elsevier; 2022, 207–237.
41. Calina D, Docea A, Petrakis D, et al. Towards effective COVID-19 vaccines: Updates, perspectives, and challenges (review). *Int J Mol Med.* 2020;46(1):3–16.
42. Ahmad RW, Salah K, Jayaraman R, Yaqoob I, Ellahham S, Omar M. Blockchain and COVID-19 pandemic: Applications and challenges. *Clust Comput.* 2023;26:1–26.
43. Ramasamy SV, Titchener-Hooker NJ, Lettieri P. Life cycle assessment as a tool to support decision making in the biopharmaceutical industry: Considerations and challenges. *Food Bioprod. Process.* 94:297–305.
44. Barbaroux M. New Plastics Economy: Rethink Biopharma SUS - BioProcess International. *BioProcess International.* [Date unknown]. Available from: https://bioprocessintl.com/manufacturing/single-use/the-green-imperative-part-two-engineering-for-the-new-plastics-economy-and-sustainability-in-single-use-technologies/
45. Whitford B, Jones D, Kinnane S. Biomanufacturing design: Reducing the environmental burden. *Curr Opin Biotechnol.* 2022;76:102717.

46. da Silva FS, da Costa CA, Crovato CD, da Rosa Righi R. Looking at energy through the lens of Industry 4.0: A systematic literature review of concerns and challenges. *Comput Ind Eng.* 2020;143:106426.

47. Mandal S, Senthil Kumar S, Singh Purushottam Kumar, Mishra SK, Harish Bishwakarma Choudhry NP, et al. Performance investigation of CuO-paraffin wax nanocomposite in solar water heater during night. *Thermochim Acta.* 2019;671:36–42.

48. Wang Y, Liu X, Han X, Godin R, Chen J, Zhou W, et al. Unique hole-accepting carbon-dots promoting selective carbon dioxide reduction nearly 100% to methanol by pure water. *Nat Commun.* 2020;11(1):2531. Wu, Z, Bañuelos, GS, Lin, Z-Q, Liu, Y, Yuan, L, Yin, X, Li, M. Biofortification and phytoremediation of selenium in China. *Front Plant Sci.* 2015;6:136.

49. Roles J, Yarnold J, Hussey K, Hankamer B. Techno-economic evaluation of microalgae high-density liquid fuel production at 12 international locations. *Biotechnol Biofuels.* 2021;14(1):1–19.

50. Nic Lughadha EM, Anna H, Belyaeva I, Nicolson N. Counting counts: Revised estimates of numbers of accepted species of flowering plants, seed plants, vascular plants and land plants with a review of other recent estimates. *Phytotaxa.* 2016;272(1):82–88.

51. Taheri S, Naimi B, Rahbek C, Araújo MB. Improvements in reports of species redistribution under climate change are required. *Sci Adv.* 2021;7(15):eabe1110.

52. Negrutiu I, Frohlich MW, Hamant O. Flowering plants in the anthropocene: A political agenda. *Trends Plant Sci.* 2020;25(4):349–368.

53. Leong YK, Chew KW, Chen W-H, Chang J-S, Show PL. Reuniting the biogeochemistry of algae for a low-carbon circular bioeconomy. *Trends Plant Sci.* 2021;26(7):729–740.

54. Zandalinas SI, Fritschi FB, Mittler R. Global warming, climate change, and environmental pollution: Recipe for a multifactorial stress combination disaster. *Trends Plant Sci.* 2021;26(6):588–599.

55. Sørensen M, Andersen-Ranberg J, Hankamer B, Møller BL. Circular biomanufacturing through harvesting solar energy and CO2. *Trends Plant Sci.* 2022;27(7):655–673.

56. Srinivasan G, Morgan D, Varun D, Brookhouser N, Brafman DA. An integrated biomanufacturing platform for the large-scale expansion and neuronal differentiation of human pluripotent stem cell-derived neural progenitor cells. *Acta Biomater.* 2018;74:168–179.

57. Groll J, Boland T, Blunk T, et al. Biofabrication: Reappraising the definition of an evolving field. *Biofabrication.* 2016;8(1):013001.

58. Kinney MA, Sargent CY, McDevitt TC. The multiparametric effects of hydrodynamic environments on stem cell culture. *Tissue Eng Part B Rev.* 2011;17(4):249–262.

59. Xu Y, Chen C, Hellwarth PB, Bao X. Biomaterials for stem cell engineering and biomanufacturing. *Bioact Mater.* 2019;4:366–379.

60. Mitsuishi M, Cao J, Bártolo P, Friedrich D, Shih AJ, Rajurkar K, Sugita N, Harada K. Biomanufacturing. *CIRP Ann.* 2013;62(2):585–606.

13 Biofabrication
Artificial Organs and Soft Tissues

Hridwin Vishaal, Sankalp Gour,
Deepak Kumar and R.K. Dwivedi

13.1 INTRODUCTION

To create a synthetic matrix that can replicate the cell microenvironment, biopolymer materials have been used. This artificial matrix needs to have the right physical and biological properties (such as orientation, topography, roughness, and stiffness), as well as physiochemical cues (such as proteins, growth factors, and signaling) to support the native ability of cells to adhere, migrate, proliferate, and differentiate towards the growth of new tissue [1]. Even though the science of biomaterials has made great strides, research and development of more efficient biomaterials for biological systems are still required. Much focus has been placed on polymer-based biomaterials in particular, due to their inherent biological and physicochemical properties. The most significant advantages of biopolymers are their excellent processability in the majority of situations, biocompatibility, stabilizability, good mechanical and physical properties, and manufacturability [2].

In the biomedical industry, natural polymers that have been derived from biological systems such as plants, animals, microbes, and algae have long been used. These substances exhibit extracellular matrix (ECM)-like structures and maintain the molecular cues and characteristics required to increase their biocompatibility [3–6]. Although in some cases, batch fluctuation in production and refining procedures limit these capacities, they often show robust cellular attachment, inhibit immunological reactions, and increase cellular function. Among the natural polymers most frequently utilized in biological applications are polysaccharides (e.g., proteins such as collagen [12], silk [13,14], gelatin [15–17], and fibrin [18]; bacterial polyesters such as bacterial cellulose [19]; alginate [6–8]; hyaluronic acid [4,9]; and chitosan [10,11]) and proteins. However, manipulation and biofabrication are sometimes difficult due to the low mechanical strength of natural polymers. To obtain the requisite mechanical properties for their usage, it is frequently essential to produce derivatives or blends using various polymers. An illustration of this is the alteration of gelatin with methacrylamide to form a possible biomaterial that might be employed in 3D bioprinting and microfluidics [20–23].

DOI: 10.1201/9781003375098-13

Complex polymeric materials that can replicate the actual ECM and restore missing or damaged tissues are needed to address actual biological challenges [24–26]. Recent improvements in biofabrication methods have made it possible to create a polymer matrix with physicochemical and compositional characteristics resembling those of the ECM. This matrix can grow and develop into proper tissue when coupled with various cell lines. Cell growth, proliferation, and transformation may also be improved by adding alternative microenvironments or other biomolecules [4,27]. Many polymeric biofabrication research lines now utilize one of two different cell inclusion techniques: (i) synthesis of a polymer matrix including encapsulated cells or (ii) cell implants on the polymer matrices that have previously been generated. Only in the last ten years has the first procedure—which is restricted to cell implantation—been used. The physical properties of the polymeric matrix material, including its hydrophobic nature, degradability, and stiffness, often determine how well these systems regenerate tissue [15,28].

These systems typically show weak cell–polymer matrix integration. Melt molding [29,30], layer by layer [30, 31], self-assembling [32], and photolithography [33] are the most frequently employed processes. The second method has attracted the greatest attention in recent years due to its capacity to create intricate structures made of several cells with intricate cellular microenvironments. Microfluidics [34,35], electrospinning [11,36], and 3D bioprinting [37,38] are recent examples of cutting-edge techniques that allow the direct integration of cells into polymer matrices with the necessary physical and biological properties to simulate the ECM of the target tissue.

This review focuses on electrospinning and 3D bioprinting, and is based on innovative polymeric matrix microfluidic biofabrication methods. The review shows that these techniques have recently been investigated for the production of polymeric matrices with encapsulated cells for biological applications.

13.2 MICROFLUIDICS

Microfluidics is a technology that examines fluids in superficial volumes by pumping fluid through a network of tubes that are as thin as a human hair and have micro to nanoscale dimensions. The chips that comprise these tubes or channels are known as microfluidic devices. The microenvironment, or the tiny environment around a cell or tissue, is an essential component to consider when creating artificial organs or tissues. The environment includes physical, biological, and physicochemical elements that collectively have an impact on the cells.

Microfluidics is utilized to create an environment that is analogous to the body's ECM, a network that surrounds cells and tissues to provide support and structure. Microgels, which are utilized to replicate the ECM, the cellular milieu required for the cells, are produced using microfluidics. Using a droplet-based microfluidic device, which forms droplets when water and oil combine, polymer droplets are produced. Cross-focusing, flow-focusing, and T-junction are three different microfluidic devices used to create droplets, as depicted in Figure 13.1. Thus, the main reasons why microgels are employed are their improved cell–matrix connections and their ability to efficiently transport mass [39,40]. The cells, ECM, and biomolecules are

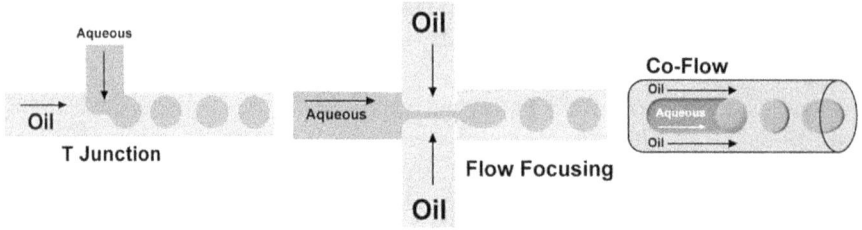

FIGURE 13.1 Schematic representation of different types of droplet-generation methods.

TABLE 13.1

An Overview of the Investigations into the Production of Cell-Filled Microgels Using Microfluidics [41–44]

Cells	Polymer	Microfluidics approach	Cross-link strategy	Microgel size range (μm)
MSCs	Alginate	Flow focusing	Ionic cross-linking	10–50
Human dermal fibroblasts	Acrylamide hyaluronic acid	Flow focusing	Enzymatic cross-linking and photopolymerization	≈80
Hepatocytes and endothelial cells	Alginate/collagen	Double emulsion (w/o/w) flow focusing	Ionic cross-linking	≤20 <0
ECFCs breast cancer cells iPSCs	GelMA/PEGDA	T-junction	Photopolymerization	300–1100
bMSCs	Thiolated gelatin sulfonated hyaluronic acid	T-junction	Thiol-Michael addition reaction	100–250
bMSCs	GelMA	Capillary	Photopolymerization	≈165

mixed to create droplets, and then the flow rates of the aqueous solution and oil are adjusted and controlled to create microgels with the necessary chemical characteristics. The cross-linking technique is selected based on the requirements for tissue encapsulation, the polymer employed, and the biomedical application. Both physical and chemical cross-linking techniques exist, and they can be used either inside a microfluidic device or after the microgels have been created [41,42]. A combination of these cross-linking techniques can be utilized, depending on the intended result. It is clear that microfluidics uses a variety of technologies and methods to create cell-filled microgels with the desired properties and parameters utilizing a variety of other polymers or polymer blends and a variety of cross-linking techniques [43,44]. Microgels are the fundamental building blocks used in the microfluidics approach to biofabrication to create replicas of organs and tissues (Table 13.1).

13.3   BIO-ELECTROSPRAYING (BES) AND CELL ELECTROSPINNING (CE)

An established technique called electrospinning is used to create nano or microfibers from substances such as natural polymers. A high-voltage power source, a collector, a nozzle, and a pump that contains the polymer solution are all components of the electrospinning process. An ultra-fine fiber is created by deforming a conical droplet of polymer solution expelled from the pump's nozzle tip into electric charges. Therefore, the polymeric solution is drawn into a cone form by the electric force generated by the electric field when a high voltage is applied to the solution's needle tip [44–46].

A liquid jet emerges from the needle tip and lands on the collector when the electric force is greater than the liquid's surface tension. The electric field used in the procedure causes the polymer to expand as well. Due to evaporation, which is brought about by high vapor pressure, the jet's diameter reduces. Additionally, the jet's diameter will continue to decrease. Thus, on the collector, nano- and microfibers are generated. Cells are positioned on mats made of the gathered nanofibers. The advantages of electrospinning technique include its affordability and straightforward scale-up process, suitability for cell culture due to nanofibers' greater surface areas, and capacity to carry out a variety of cell functions [47,48]. Since traditional electrospinning has significant drawbacks that result in uneven cell distribution, cell electrospinning is a comparable technology that is utilized in place of electrospinning. In cell electrospinning, living cells and a polymer solution are mixed to create scaffolds, which enhance cell distribution.

The extremely strong electric force applied to a jet leads it to split into nanoparticles because of the lower viscosity of the polymeric material used. The same machine that is used for electrospinning is also used for electrospraying. The viscosity of the polymer is the only variable thus far. Similar methods include bio-electrospraying, in which living cells are encapsulated in the resultant scaffolds by a polymer solution. It should be noted that the electrospinning technique does not have an impact on the population or health of the cells used [49,50]. The material characteristics, such as the solution's viscosity, and the mechanism parameters, such as the applied electric field and collector distance, must be controlled for the encapsulated cells to have good vitality and the fibers to have the necessary form and size. Typically, only electrospinning and electrospraying processes make use of synthetic polymers. Natural polymers are frequently used in cell electrospinning and biospraying techniques for biofabrication, which have advantages such as a high cell affinity and ECM biometric features because they are naturally generated and hence better suited for living cells [51,52].

The low mechanical strength of the scaffold, however, is a key disadvantage of these methods. The mechanism can be improved, and a range of natural and artificial polymer combinations can be employed for the polymer solution to increase the mechanical strength. Table 13.2 displays the various outcomes. The data lead to the conclusion that the experiment's settings significantly affect cell viability. This technology, combined with 3D bioprinting, can be used to create complicated scaffold structures while improving their mechanical strength and characteristics [53,54].

TABLE 13.2

**Summary of Research on Cell Electrospinning and Bio-Electrospraying
[44–56]**

Cells	Blend	Electric field	Cell viability (%)
MG63	Alginate/poly(ethylene oxide)/ lecithin	0.16 kV/mm	80
C2C12 myoblast	Alginate/PEO	0.075 kV/mm	90
HUVECs	Alginate/PEO	10.5 kV/mm	90
Adipose derived from ADSCs	Gelatin/pullulan	8 kV	90
C2C12 murine myoblast	Fibrin/PEO	4.5 kV	7 days
Primary cardiomyocytes	Matrigel/laminin	230 V	80

13.4 3D BIOPRINTING

Bioprinting is a widely used technique that allows for the printing of objects from a design produced by software. It is possible to build cell-free scaffolds using traditional 3D bioprinting. As a result, it makes way for the printing of scaffold structures to construct complex functional living tissues. The capacity to print small, low-volume scaffold constructions that can be utilized for mending or building tissues and even organs, as well as 3D bioprinting's cost-effectiveness, are two benefits.

The printing bio-ink utilized in the procedure is a crucial component. There have been recent developments in the creation of bio-inks devoid of polymers, which only contain cells and are called biocompatible bio-inks since polymer-based bio-inks frequently have toxic qualities inappropriate for cells. The cytotoxicity of each substance contained in the substrate must also be taken into account. With the relatively recent development of 4D printing in bioprinting, it is now possible to create intricate three-dimensional constructions of tissues and organs that may change shape in response to stimuli from the outside world [57,58]. Living cells and cell structures created using 4D bioprinting have the potential to change and mature over time. The main problem with biosensors is created by utilizing 4D printing that can react to minute metabolites. Inkjet, extrusion, laser-assisted 3D bioprinting, and stereolithography are some of the main methods employed. Figure 13.2 describes each technique and Table 13.3 provides a summary of the techniques.

The main issue with using bio-inks is their lack of mechanical strength and low cell affinity, thus several derivatives that can promote cell growth, adhesion, and differentiation as well as improve mechanical strength and cell affinity are being created. Alginate hydrogel–based bio-inks are utilized because of their excellent biocompatibility and resemblance to the ECM, but their poor mechanical strength makes them difficult to print. Depending on the ratio of the material composition, alginate is cross-linked with other materials to improve the properties of the bio-ink, such as printability and cell survival. Another natural polymer that can be utilized in biofabrication is chitosan. It has qualities such as biodegradability and antibacterial activity

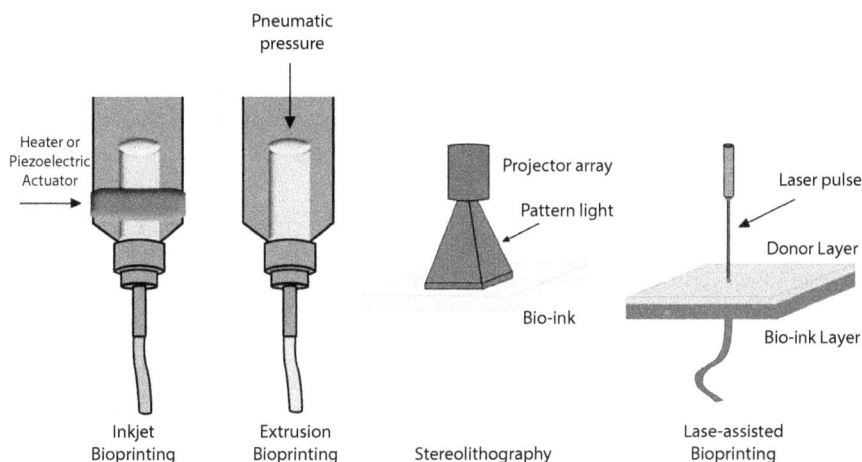

FIGURE 13.2 Schematic representation of various bioprinting techniques.

TABLE 13.3
Summary of Research on 3D Printing Technologies [44–56]

Technique	Operation	Properties
Inkjet bioprinting	Droplets are extruded from a polymer solution using a piezoelectric actuator, a thermal actuator, and electrostatic forces	Only applicable to solutions with low viscosity and low cell density
Extrusion-based bioprinting	Staking of layers of bio-ink by the extrusion of a polymer through a micro-nozzle by applying continued pressure	Capacity to print highly viscous bio-inks with high density and high cell viability
Laser-assisted bioprinting	Laser is used as an energy source to deposit biomaterials onto a substrate	High degree of precision and resolution with highly viscous bio-inks with high cell density
Stereolithography	Layer by layer, photolytic cross-linking of bio-inks is accomplished with light	Greatest possible resolution and highest biological properties achievable

and is more compatible with living tissues [59–61]. Chitosan is chemically cross-linked with substances such as genipin and glutaraldehyde to improve the printability, accuracy, and strength of the resultant scaffold. Sacrificial bio-inks in the bioprinting process are used to mechanically support the scaffold while it is being printed. Sacrificial bio-ink is made of a low-adhesion material that is removed when the scaffold has been printed, with the scaffold keeping its structural integrity. Directly printing tissue and organs is one of the uses of 3D bioprinting in biochemistry [62,63].

Advancements are being made in the creation of artificial tissues that replicate the actions and capabilities of biological tissues for the replacement of organs. The creation of tissues with tissue growth factors, dietary components, and their physiological activities is a recent emphasis of 3D bioprinting. Since the majority of tissues in the body are vascularized, the vascularization of engineered tissues affects the success of bioprinted tissues, but its absence is one of the main reasons why bioprinted tissues fail. Combining cells, proteins, and scaffolds has been utilized in a variety of ways to create larger 3D tissues and organs from cell sheets. Most researchers have been successful in creating a vascularized, perfusable human skin. It is fully functional, has a microenvironment similar to that of actual skin, and can be used to test skin medications [64,65].

13.5 CONCLUSION

This review examined the interaction between innovative biofabrication methods and natural polymers. Much interest has been shown in the biofabrication of naturally occurring polymer scaffolding with encapsulated cells employing microfluidics, electrospinning, and 3D bioprinting techniques. There are many challenges in successfully biofabricating these cell-filled natural scaffolds. In this regard, many alterations and mixtures have been researched to enhance the processed scaffold's mechanical qualities. On the one side, the weak mechanical strength of naturally occurring polymers makes production and material manipulation difficult. Processing must not affect the encapsulated cells' viability or effectiveness. Hence, care must be taken while using these approaches.

High cell survival, metabolic activity, and the proliferation of the encapsulated cells are produced by controlling both material properties (such as feed rate, voltage, or pressure) and material properties (such as polymer content, viscosity, or solvent). It is consequently regarded as the most critical and needs to be dealt with right away. Even though several modifications, blends, and alterations of the biofabrication procedure have been investigated, we are still in the early stages of creating these technologies. Many technical issues and limitations still need to be resolved. One of the most pressing issues is how to use the cell-filled scaffolds created using these techniques for medicinal purposes. Given the numerous efforts being made to develop more sophisticated techniques to replicate the intricacies of native tissues and overcome manufacturing limitations, the possibility cannot be ruled out that in a few years, biofabrication techniques will advance and enable the creation of fully functional organs and tissues.

REFERENCES

1. Tutar, R., Motealleh, A., Khademhosseini, A., & Kehr, N. S. (2019). Functional nanomaterials on 2D surfaces and in 3D nanocomposite hydrogels for biomedical applications. *Advanced Functional Materials*, 29(46), 1904344.
2. Samadian, H., Maleki, H., Allahyari, Z., & Jaymand, M. (2020). Natural polymers-based light-induced hydrogels: Promising biomaterials for biomedical applications. *Coordination Chemistry Reviews*, 420, 213432.

3. Asadi, N., Del Bakhshayesh, A. R., Davaran, S., & Akbarzadeh, A. (2020). Common biocompatible polymeric materials for tissue engineering and regenerative medicine. *Materials Chemistry and Physics*, 242, 122528.

4. Mora-Boza, A., Puertas-Bartolomé, M., Vázquez-Lasa, B., San Román, J., Perez-Caballer, A., & Olmeda-Lozano, M. (2017). Contribution of bioactive hyaluronic acid and gelatin to regenerative medicine. Methodologies of gels preparation and advanced applications.

5. Nolan, K., Millet, Y., Ricordi, C., & Stabler, C. L. (2008). Article commentary: Tissue engineering and biomaterials in regenerative medicine. *Cell Transplantation*, 17(3), 241–243.

6. Eiselt, P., Yeh, J., Latvala, R. K., Shea, L. D., & Mooney, D. J. (2000). Porous carriers for biomedical applications based on alginate hydrogels. *Biomaterials*, 21(19), 1921–1927.

7. Yeo, M., & Kim, G. (2015). Fabrication of cell-laden electrospun hybrid scaffolds of alginate-based bioink and PCL microstructures for tissue regeneration. *Chemical Engineering Journal*, 275, 27–35.

8. Cook, M. T., Tzortzis, G., Charalampopoulos, D., & Khutoryanskiy, V. V. (2011). Production and evaluation of dry alginate-chitosan microcapsules as an enteric delivery vehicle for probiotic bacteria. *Biomacromolecules*, 12(7), 2834–2840.

9. Sideris, E., Griffin, D. R., Ding, Y., Li, S., Weaver, W. M., Di Carlo, D., … Segura, T. (2016). Particle hydrogels based on hyaluronic acid building blocks. *ACS Biomaterials Science and Engineering*, 2(11), 2034–2041.

10. Husain, S., Al-Samadani, K. H., Najeeb, S., Zafar, M. S., Khurshid, Z., Zohaib, S., & Qasim, S. B. (2017). Chitosan biomaterials for current and potential dental applications. *Materials*, 10(6), 602.

11. Qasim, S. B., Zafar, M. S., Najeeb, S., Khurshid, Z., Shah, A. H., Husain, S., & Rehman, I. U. (2018). Electrospinning of chitosan-based solutions for tissue engineering and regenerative medicine. *International Journal of Molecular Sciences*, 19(2), 407.

12. Chan, H. F., Zhang, Y., & Leong, K. W. (2016). Efficient one-step production of microencapsulated hepatocyte spheroids with enhanced functions. *Small*, 12(20), 2720–2730.

13. Zafar, M. S., & Al-Samadani, K. H.. Potential use of natural silk for bio-dental applications. *Journal of Taibah University Medical Sciences*, 9(3), 171–177.

14. Zafar, M. S., Belton, D. J., Hanby, B., Kaplan, D. L., & Perry, C. C. (2015). Functional material features of Bombyx mori silk light versus heavy chain proteins. *Biomacromolecules*, 16(2), 606–614.

15. Yang, C. Y., Chiu, C. T., Chang, Y. P., & Wang, Y. J. (2009). Fabrication of porous gelatin microfibers using an aqueous wet spinning process. *Artificial Cells, Blood Substitutes, and Biotechnology*, 37(4), 173–176.

16. Cha, C., Oh, J., Kim, K., Qiu, Y., Joh, M., Shin, S. R., … Khademhosseini, A. (2014). Microfluidics-assisted fabrication of gelatin-silica core–shell microgels for injectable tissue constructs. *Biomacromolecules*, 15(1), 283–290.

17. Feng, Q., Li, Q., Wen, H., Chen, J., Liang, M., Huang, H., … & Cao, X. (2019). Injection and self-assembly of bioinspired stem cell-laden gelatin/hyaluronic acid hybrid microgels promote cartilage repair in vivo. *Advanced Functional Materials*, 29(50), 1906690.

18. Guo, Y., Gilbert-Honick, J., Somers, S. M., Mao, H. Q., & Grayson, W. L. (2019). Modified cell-electrospinning for 3D myogenesis of C2C12s in aligned fibrin microfiber bundles. *Biochemical and Biophysical Research Communications*, 516(2), 558–564.

19. Jayani, T., Sanjeev, B., Marimuthu, S., & Uthandi, S. (2020). Bacterial Cellulose Nano Fiber (BCNF) as carrier support for the immobilization of probiotic, Lactobacillus acidophilus 016. *Carbohydrate Polymers*, 250, 116965.

20. Aldana, A. A., Valente, F., Dilley, R., & Doyle, B. (2021). Development of 3D bio-printed GelMA-alginate hydrogels with tunable mechanical properties. *Bioprinting*, 21, e00105.

21. Lee, D., Lee, K., & Cha, C. (2018). Microfluidics-assisted fabrication of microtissues with tunable physical properties for developing an in vitro multiplex tissue model. *Advanced Biosystems*, 2(12), 1800236.

22. Seeto, W. J., Tian, Y., Pradhan, S., Kerscher, P., & Lipke, E. A. (2019). Rapid production of cell-laden microspheres using a flexible microfluidic encapsulation platform. *Small*, 15(47), 1902058.

23. Li, F., Truong, V. X., Thissen, H., Frith, J. E., & Forsythe, J. S. (2017). Microfluidic encapsulation of human mesenchymal stem cells for articular cartilage tissue regeneration. *ACS Applied Materials and Interfaces*, 9(10), 8589–8601.

24. Lee, J., Cuddihy, M. J., & Kotov, N. A. (2008). Three-dimensional cell culture matrices: State of the art. *Tissue Engineering. Part B: Reviews*, 14(1), 61–86.

25. Fabbri, M., Garcia-Fernandez, L., Vazquez-Lasa, B., Soccio, M., Lotti, N., Gamberini, R., ... San Roman, J. (2017). Micro-structured 3D-electrospun scaffolds of biodegradable block copolymers for soft tissue regeneration. *European Polymer Journal*, 94, 33–42.

26. Ribeiro, V. P., Pina, S., Costa, J. B., Cengiz, I. F., García-Fernández, L., Fernández-Gutiérrez, M. D. M., ... Reis, R. L. (2019). Enzymatically cross-linked silk fibroin-based hierarchical scaffolds for osteochondral regeneration. *ACS Applied Materials and Interfaces*, 11(4), 3781–3799.

27. Puertas-Bartolomé, M., Benito-Garzon, L., Fung, S., Kohn, J., Vázquez-Lasa, B., & San Román, J. (2019). Bioadhesive functional hydrogels: Controlled release of catechol species with antioxidant and antiinflammatory behavior. *Materials Science and Engineering: C*, 105, 110040.

28. Eltom, A., Zhong, G., & Muhammad, A. (2019). Scaffold techniques and designs in tissue engineering functions and purposes: A review. *Advances in Materials Science and Engineering*, 2019.

29. Fu, J., Li, X. B., Wang, L. X., Lv, X. H., Lu, Z., Wang, F., ... Li, C. M. (2020). One-step dip-coating-fabricated core–shell silk fibroin rice paper fibrous scaffolds for 3D tumor spheroid formation. *ACS Applied Bio Materials*, 3(11), 7462–7471.

30. Tsukamoto, Y., Akagi, T., & Akashi, M. (2020). Vascularized cardiac tissue construction with orientation by layer-by-layer method and 3D printer. *Scientific Reports*, 10(1), 1–11.

31. Siddiq, A., & Kennedy, A. R. (2020). Compression moulding and injection over moulding of porous PEEK components. *Journal of the Mechanical Behavior of Biomedical Materials*, 111, 103996.

32. van Bochove, B., & Grijpma, D. W. (2021). Mechanical properties of porous photocrosslinked poly (trimethylene carbonate) network films. *European Polymer Journal*, 143, 110223.

33. Peck, M., Dusserre, N., McAllister, T. N., & L'Heureux, N. (2011). Tissue engineering by self-assembly. *Materials Today*, 14(5), 218–224.

34. Chen, Q., Chen, D., Wu, J., & Lin, J. M. (2016). Flexible control of cellular encapsulation, permeability, and release in a droplet-templated bifunctional copolymer scaffold. *Biomicrofluidics*, 10(6), 064115.

35. Rossow, T., Lienemann, P. S., & Mooney, D. J. (2017). Cell microencapsulation by droplet microfluidic templating. *Macromolecular Chemistry and Physics*, 218(2), 1600380.

36. Zafar, M., Najeeb, S., Khurshid, Z., Vazirzadeh, M., Zohaib, S., Najeeb, B., & Sefat, F. (2016). Potential of electrospun nanofibers for biomedical and dental applications. *Materials*, 9(2), 73.

37. Zimmerling, A., Yazdanpanah, Z., Cooper, D. M., Johnston, J. D., & Chen, X. (2021). 3D Bioprinting PCL/nHA bone scaffolds: Exploring the influence of material synthesis techniques. *Biomaterials Research*, 25(1), 1–12.

38. Vyas, C., Zhang, J., Øvrebø, Ø., Huang, B., Roberts, I., Setty, M., ... Bartolo, P. (2021). 3D Bioprinting of silk microparticle reinforced polycaprolactone scaffolds for tissue engineering applications. *Materials Science and Engineering: C*, 118, 111433.

39. Jiang, W., Li, M., Chen, Z., & Leong, K. W. (2016). Cell-laden microfluidic microgels for tissue regeneration. *Lab on a Chip*, 16(23), 4482–4506.

40. Agrawal, G., & Agrawal, R. (2018). Functional microgels: Recent advances in their biomedical applications. *Small*, 14(39), 1801724.

41. Utech, S., Prodanovic, R., Mao, A. S., Ostafe, R., Mooney, D. J., & Weitz, D. A. (2015). Microfluidic generation of monodisperse, structurally homogeneous alginate microgels for cell encapsulation and 3D cell culture. *Advanced Healthcare Materials*, 4(11), 1628–1633.

42. Mao, A. S., Shin, J. W., Utech, S., Wang, H., Uzun, O., Li, W., ... Mooney, D. J. (2017). Deterministic encapsulation of single cells in thin tunable microgels for niche modelling and therapeutic delivery. *Nature Materials*, 16(2), 236–243.

43. Ma, T., Gao, X., Dong, H., He, H., & Cao, X. (2017). High-throughput generation of hyaluronic acid microgels via microfluidics-assisted enzymatic crosslinking and/or Diels–Alder click chemistry for cell encapsulation and delivery. *Applied Materials Today*, 9, 49–59.

44. Cheng, Y., Zhang, X., Cao, Y., Tian, C., Li, Y., Wang, M., ... Zhao, G. (2018). Centrifugal microfluidics for ultra-rapid fabrication of versatile hydrogel microcarriers. *Applied Materials Today*, 13, 116–125.

45. Doshi, J., & Reneker, D. H. (1995). Electrospinning process and applications of electrospun fibers. *Journal of Electrostatics*, 35(2–3), 151–160.

46. Agarwal, S., Wendorff, J. H., & Greiner, A. (2008). Use of electrospinning technique for biomedical applications. *Polymer*, 49(26), 5603–5621.

47. Zanin, M. H. A., Cerize, N. N., & Oliveira, A. M. D. (2011). Production of nanofibers by electrospinning technology: Overview and application in cosmetics. *Nanocosmetics and Nanomedicines*, 50(23), 311–332.

48. De Lima, G. G., Lyons, S., Devine, D. M., & Nugent, M. J. (2018). Electrospinning of hydrogels for biomedical applications. In: Thakur, V. K., Thakur, M. K. (eds) *Hydrogels* (pp. 219–258). Singapore: Springer.

49. Bhardwaj, N., & Kundu, S. C. (2010). Electrospinning: A fascinating fiber fabrication technique. *Biotechnology Advances*, 28(3), 325–347.

50. Bhushani, J. A., & Anandharamakrishnan, C. (2014). Electrospinning and electrospraying techniques: Potential food based applications. *Trends in Food Science and Technology*, 38(1), 21–33.

51. Bock, N., Woodruff, M. A., Hutmacher, D. W., & Dargaville, T. R. (2011). Electrospraying, a reproducible method for production of polymeric microspheres for biomedical applications. *Polymers*, 3(1), 131–149.

52. Greig, D., & Jayasinghe, S. N. (2008). Genomic, genetic and physiological effects of bioelectrospraying on live cells of the model yeast Saccharomyces cerevisiae. *Biomedical Materials*, 3(3), 034125.

53. Mongkoldhumrongkul, N., Swain, S. C., Jayasinghe, S. N., & Stürzenbaum, S. (2010). Bio-electrospraying the nematode Caenorhabditis elegans: Studying whole-genome transcriptional responses and key life cycle parameters. *Journal of the Royal Society. Interface*, 7(45), 595–601.

54. Poncelet, D., de Vos, P., Suter, N., & Jayasinghe, S. N. (2012). Bio-electrospraying and cell electrospinning: Progress and opportunities for basic biology and clinical sciences. *Advanced Healthcare Materials*, 1(1), 27–34.

55. Vatankhah, E., Prabhakaran, M. P., & Ramakrishna, S. (2015). Biomimetic nanostructures by electrospinning and electrospraying. In H. Baharvand & N. Aghdami (eds), *Stem-Cell Nanoengineering*, 123–141. John Wiley & Sons, Inc.

56. Jeong, S. B., Chong, E. S., Heo, K. J., Lee, G. W., Kim, H. J., & Lee, B. U. (2019). Electrospray patterning of yeast cells for applications in alcoholic fermentation. *Scientific Reports*, 9(1), 1–7.

57. Tycova, A., Prikryl, J., Kotzianova, A., Datinska, V., Velebny, V., & Foret, F. (2021). Electrospray: More than just an ionization source. *Electrophoresis*, 42(1–2), 103–121.

58. Nesaei, S., Song, Y., Wang, Y., Ruan, X., Du, D., Gozen, A., & Lin, Y. (2018). Micro additive manufacturing of glucose biosensors: A feasibility study. *Analytica Chimica Acta*, 1043, 142–149.

59. Jungst, T., Smolan, W., Schacht, K., Scheibel, T., & Groll, J. (2016). Strategies and molecular design criteria for 3D printable hydrogels. *Chemical Reviews*, 116(3), 1496–1539.

60. Yue, Z., Liu, X., Coates, P. T., & Wallace, G. G. (2016). Advances in printing biomaterials and living cells: Implications for islet cell transplantation. *Current Opinion in Organ Transplantation*, 21(5), 467–475.

61. Muthukrishnan, L. (2021). Imminent antimicrobial bioink deploying cellulose, alginate, EPS and synthetic polymers for 3D bioprinting of tissue constructs. *Carbohydrate Polymers*, 260, 117774.

62. Rubio-Valle, J. F., Perez-Puyana, V., Jiménez-Rosado, M., Guerrero, A., & Romero, A. (2021). Evaluation of smart gelatin matrices for the development of scaffolds via 3D bioprinting. *Journal of the Mechanical Behavior of Biomedical Materials*, 115, 104267.

63. Ma, L., Li, Y., Wu, Y., Yu, M., Aazmi, A., Gao, L., ... Yang, H. (2020). 3D bioprinted hyaluronic acid-based cell-laden scaffold for brain microenvironment simulation. *Bio-Design and Manufacturing*, 3(3), 164–174.

64. Redmond, J., McCarthy, H., Buchanan, P., Levingstone, T. J., & Dunne, N. J. (2021). Advances in bio-fabrication techniques for collagen-based 3D in vitro culture models for breast cancer research. *Materials Science and Engineering: C*, 122, 111944.

65. Soltan, N., Ning, L., Mohabatpour, F., Papagerakis, P., & Chen, X. (2019). Printability and cell viability in bioprinting alginate dialdehyde-gelatin scaffolds. *ACS Biomaterials Science and Engineering*, 5(6), 2976–2987.

14 Capabilities and Limitations of Materials and Manufacturing Process in Medical Implants

Sudhanshu Kumar, Deepak Kumar, and Ashutosh Kumar

14.1 INTRODUCTION

One of the prime concerns of the health sector is to provide reliable and safe bioimplants to patients. The most important consideration for any medical implant is to function like natural bone of the living body. Artificial implants should be non-reactive to human tissue, non-carcinogenic, and must have sufficient strength to resist the load in the working environment of the human body [1]. Bioimplants should not react with the tissues surrounding them. A wide variety of materials are used for the purpose of bioimplants. Metals and alloys, e.g., titanium (Ti), steel, cobalt-chromium (Co-Cr), magnesium (Mg), iron (Fe), zirconium (Zr), niobium (Nb) alloy, polymers and their composites, ceramics, and nanocomposites [2–10] are mostly used in dental, orthopedic, and reconstructive surgery. Despite the availability of metals and non-metals, these materials do not provide optimal solutions due to issues such as machinability, fitting difficulties, strength and bio integration characteristics. Bone reabsorption has been observed with metallic implants having a high modulus of elasticity [11].

A higher coefficient of friction is one of the reasons for the short lifespan of moving implants. Movable bioimplants need a very high finish to prevent frictional wear. As polishing is difficult on implants due to their complex curves, a surface coating with a low friction coefficient and anti-wear is an attractive solution. A bioactive coating provides a high antibacterial effect over a long time period. Nobel metals such as platinum, palladium, silver, and gold have very high potential as bioimplant material because of their high inertness and corrosion resistance. However, their long-term effect and compatibility with other materials need more research.

DOI: 10.1201/9781003375098-14

The manufacturing of bioimplants is another aspect of research and development in this area. Owing to the complexity and functionality of biomaterials, manufacturing techniques are a subject for research and development in fulfilling the requirements of implants. The manufacturing of bioimplants involves a number of steps such as computer-aided design (CAD), simulations, machining as per biomaterials, finishing, and fixation [12]. The manufacturing techniques of implants depend on the type of implant. For example, screws, plates, and prostheses, the most common types of orthopedic implants, are in high demand requiring mass manufacturing. The primary machining of a metallic block starts with computer numerical control (CNC) milling and turning machines. Stock removal and precise turning, cutting, and drilling are common operations carried out by multi-axis milling machines with robotic arms. Intricate cuts on implants are performed by the electrical discharge machining (EDM) process. A matt texture on the surface of implants can be created by sand blasting. Additive manufacturing (AM), also known as 3D printing, is one of the manufacturing techniques that is emerging as a reliable process for manufacturing metallic biomaterials. In this process, the raw material is melted by a laser beam or an electron beam source and deposited layer by layer according to CAD. The additive manufacturing process is mainly used for the development of porous structured implants, and is widely used to develop dental parts [13], trauma medical implants, and orthopedic medical devices [14]. Since medical implants are highly customizable, single point incremental forming (SPIF) is highly applicable. Incremental forming does not require a preshaped die and punch and it is very flexible. Titanium alloy, stainless steel, and magnesium alloy sheets can be easily deformed into the desired shape using SPIF. Casting is also a popular manufacturing method that is used for the fabrication of implants made of a metal matrix composite. This chapter discusses biocompatible materials, applications, capabilities, and their limitations and the manufacturing techniques used for various types of bioimplants.

14.2 BIOCOMPATIBLE MATERIALS

Biocompatible materials do not produce any toxicity when in contact with tissue. They should be non-inflammatory and non-reactive, and should function in an appropriate manner with respect to the natural body parts. Biocompatible materials include Ti and its alloys, which are the most common material for bioimplants, ceramics, and polymers.

14.2.1 Titanium-Based Bioimplants

Ti and its alloys are one of the most applicable materials for medical implants due to their high biocompatibility, high corrosion resistance, and high strength with exceptional plastic characteristics [15]. Despite these properties, Ti alloys have low bioactivity and a high coefficient of friction causing the implants to fail during use. Several efforts have been made to improve the life of implants as well as their biofunctionality. One such effort is the surface modification and material composition of an implant's surface. In order to increase the antibacterial effect with high wear resistance, the coating of nitrates TiNx on the Ti surface has attracted attention [8–13]. Wu at el. [16] studied the effect of Cu-N coating on titanium implants with respect to wear and antibacterial behavior. The coating was created using a duplex surface treatment, i.e., magnetron sputtering and plasma nitriding. It was observed that the coating of Cu-N on a titanium alloy improves the antibacterial property along with wear and corrosion resistance. Antibacterial properties due to the fusion of a copper element with Ti were observed by Zhang et al. [15]. The copper element was added using the powder metallurgy method. The Ti-Cu alloy has shown antibacterial properties. However, antibacterial properties are found when bacteria comes in contact with Cu. Porous-structured implants are suitable for tissue growth similar to natural bones. Porous metal implants can be made from nitinol (NiTi) or titanium nickelide, which has good elasticity characteristics with high strength [17].

14.2.2 POLYMER-BASED BIOIMPLANTS

Polymers are growing in popularity for medical implants due to their lightness, inertness toward biochemicals, and excellent flexibility. A wide variety of polymers such as polymethylmethacrylate (PMMA), polyethylene (PE), polyurethane (PU), and polyethylene terephthalate (PET) are computable for bioimplants. However, polymers are not as strong as metals and are also limited in their interchangeability. Polyaryletherketones (PAEKs) are widely used for creating bioimplants, especially in the fields of orthopedics, trauma, and spinal implants.

14.2.3 CERAMIC IMPLANTS

Ceramic materials are widely used in biomedical applications for their mechanical properties, biocompatibility, biodegradability, and high corrosive strength [18]. One of their major applications is in biomedical implants [19]. Clays and their mix with metal oxides, mineral fillers, and other inorganic compounds are sintered to produce ceramic materials. Due to the growing recognition that wear debris produced by metal–polyethylene components of total hip prostheses may lead to osteolysis around implants, there has been a significant increase in interest in ceramic implants [20]. The main purpose of a ceramic implant is to fill the spaces between bones as well as to eliminate the need for metallic implants. Depending on how the body responds to the implant, bioceramic materials can be divided into three categories: bioinert (Al_2O_3, ZrO_2, pyrolytic carbon), bioactive (glass based on Na_2O-CaO-P_2O_5-SiO_2), and completely biodegradable (hydroxyapatite $Ca10(–PO_4)_6(OH)_2$ and tricalcium phosphate $Ca_3(PO_4)_2$) [21,22].

Bioactive ceramics are set apart from other types of ceramics by their capacity to biodegrade and release ions that establish a chemical bond between the implant and the bone's mineralization, ultimately promoting osseointegration. Therefore, by adjusting the composition of the bioactive ceramics, it is possible to modify the resorbability and bioactivity of implants [23]. Several bioactive glasses comprising specific quantities of SiO_2, CaO, and P_2O_5 were first to demonstrate bonding to bone [6]. Ceramics made of calcium phosphate with a bioresorbable composition break down hydrolytically in the body as natural tissues regenerate to take their place. In this instance, the body's metabolic mechanisms absorb and release the chemical byproducts of breakdown. The interaction between the stable apatite crystals in the bone and the glass ceramics is believed to occur via solid-state bonding, which is regarded as a chemical process and the cause of the interface reaction [24]. The cellular composition and thickness of the capsule that forms around an implant can provide insight into its biocompatibility. Bioinert ceramics, including corundum and ceramics made of zirconium dioxide or aluminum oxide due to their low coefficient of friction, resistance to corrosion, and appropriate mechanical features including high wear resistance and strength, are common alternatives for orthopedic and dental implants. Inside the body, these ceramics do not undergo any chemical changes and develop a fibrous capsule that protects against foreign bodies. The preparation techniques and ceramic synthesis conditions determine the porosity and microstructure of the implants [22].

Bioinert ceramics can be produced in a variety of ways, such as by casting thermoplastic slips to generate thin-walled objects with intricate shapes. When the standard pressing process followed by sintering cannot provide a dense product, cold pressing with additional sintering is utilized. Hot pressing is a technique that may be used to create high-strength materials from tiny powders. The plasma–chemical technique permits the synthesis of high-temperature phases, including solid solutions with the limited solubility of one component in another, due to the rapid rate of cooling of the reaction products. Traditional sintering methods produce ceramics with low-strength characteristics and coarse-grain structures [25]. Despite the fact that ceramic materials are better for the body's biochemistry, their use in bone prosthetics and implants is limited due to their low durability, which can lead to mechanical loads causing resorption of the surrounding bone tissue and implant failure. Recent incidents of zirconium prosthetic implant failures have resulted from fast crack propagation. To improve the lifespan and reliability of ceramic implants, new research has focused on developing highly crack-resistant aluminum and zirconium nanocomposites [26]. To improve the process of the osteointegration of stable or bioinert ceramics with the surrounding bone tissue, modifying their surfaces has been considered. Due to their resistance, it is crucial for the material to possess antimicrobial properties that help to stop harmful microbe adhesion and proliferation. Spraying bioactive substances on surfaces, changing surface topography, and increasing porosity in regions that will come into contact with bone tissue are a few surface modification techniques. Another approach to modifying ceramics is through composite materials that offer novel properties [27]. Due to their resistive nature, it is essential that the material has antibacterial capabilities to prevent pathogenic microbe attachment and reproduction [28]. Modification techniques include spraying bioactive components, changing the porosity and surface topography in the area that will come into contact with the bone tissue, and more. Making composite materials with novel qualities is an alternative method for modifying ceramics [29].

14.3   MANUFACTURING OF BIOIMPLANTS

Manufacturing systems have adopted their processes for the production of bioimplants. Broadly, the manufacture of bioimplants can be observed in two categories: shaping of the implants which involves casting, additive manufacturing, sintering, and forming; and finishing or cutting the surfaces which involves milling, turning, polishing, and coating. The following sections discuss the manufacturing processes.

14.3.1   ADDITIVE MANUFACTURING BIOMEDICAL IMPLANTS

The additive manufacturing technique, sometimes referred to as 3D printing, adds the material layer by layer without the need for any complicated post-processing stages, creating a complex geometric porous structure [30]. This manufacturing process is typically described as adding a feedstock material layer by layer in accordance with specified tool paths from a computer-aided design, as opposed to subtractive methods. Accurate, economical, and rapid treatments have been made possible over time using 3D printing technology. An essential component of AM is computer-aided design. By departing from conventional manufacturing techniques, which are subtractive in nature, AM makes it possible to fabricate CAD models on 3D printers with the least amount of raw material waste [31]. An implant is a medical device that is made to replace a biological structure that is missing, to maintain a biological structure that has been harmed, or to enhance a biological structure that already exists. Medical implants are manufactured objects, as opposed to transplants, which involve the implantation of biological tissue. Implant surfaces that come into contact with the body may be made of biomedical substances such as silicone, titanium, or apatite, depending on which is most practical. There are several applications for additive manufacturing. Every patient is different in the medical and dentistry fields; therefore, AM holds great promise for providing individualized and tailored treatments. In the field of dentistry, AM is used for splints, braces, dental models, and drill guides. However, the possibility of using AM to develop artificial tissues and organs has also been explored. Beginning with imaging or recording the patient's geometry using computed tomography or other 3D scanning techniques, a typical procedure for customized medical equipment is to begin with the patient. These data are then used to create a three-dimensional representation of the patient's anatomy, which may be seen as the early use of additive manufacturing for a medical model [31].

The most popular formats for 3D models required by each AM process and piece of machinery are standard triangle language, stereolithography, and standard tessellation language (STL). The STL model is then divided into layers and processed to provide instructions for the particular AM machine [32]. Medical implants are produced using a variety of additive manufacturing processes. Powder bed fusion (PBF), in particular selective laser melting (SLM), is the most used metal additive manufacturing technology. The mechanically modified Ti6Al4V bone scaffold, patient-specific hip implants, dental implants, and titanium interbody fusion cages have all been produced using the SLM method [33].

14.3.1.1 Selective Laser Melting

An AM process that can create large structural elements with essentially no geometric restrictions is SLM. The term "selective" suggests that a single or small amount of powder has been processed. The term "laser" denotes the employment of a laser for processing, while "melting" denotes the complete melting of the assigned powder. Typically, SLM processing begins with creating the final product in a secure environment using CAD software and a pre-tooled 3D model. An energy source selectively melts a metallic powder used in the SLM process, which is uniformly dispersed throughout the build platform [34]. The focus and mobility of the laser beam on the building table are controlled by a laser diffraction apparatus composed of a flatter sector lens and Galvano mirrors. SLM selectively melts the powder particles point by point, line by line, and layer by layer to print the complete near-net-shape component with 99.9% relative density.

Due to its adaptability and high precision, as well as its superior surface polish and the structural integrity of the manufactured implants, SLM can be seen as a particularly promising method for the manufacture of medical devices. SLM is a great option for producing tiny features (500 µm) that are frequently found in spinal fusion implant devices [35] (Figure 14.1).

14.3.1.2 Single Point Incremental Forming

In recent years, the desire of the growing senior population to maintain their quality of life, together with the high incidence of accidents resulting in non-fatal injuries, has contributed to an increase in the demand for personalized prostheses. Density, elastic modulus, fracture toughness, and compressive yield strength of magnesium

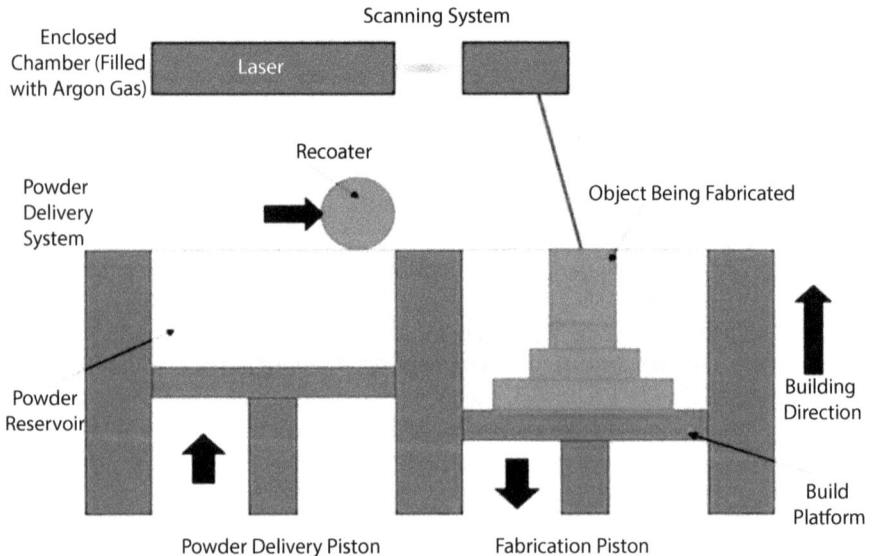

FIGURE 14.1 Typical SLM process [35].

(Mg) alloys are more comparable to those of natural bones than other metallic materials commonly used for implants, such as stainless steels, titanium alloys, and magnesium (Mg) alloys. [37]. Using standard computer numerical control machines, SPIF technology enables the production of highly customizable goods. Magnesium is a metal that is both biocompatible and biodegradable, and it has mechanical qualities that are similar to those of cortical bones. This makes it a good option for temporary biomedical prostheses and eliminating the need for additional surgery to remove them [38].

SPIF is a particularly adaptable technology since it doesn't require specialized equipment or unique dies, which also lowers costs and setup times. Additionally, rapid modifications to the part's design can made simply by altering the spinning punch's direction. Due to the use of the incremental forming process, which carefully adhered to technological limits, a specific procedure for the best placing of the shape into the forming frame has been carried out.. The support is divided into two complementary portions that may be created independently to achieve a simpler design. Using a precise assessed technique (laser, water jet), the individual shells can be separated from the respective native sheet and the two resulting pieces can then be joined to form the final support using conventional banding techniques.

The differences, determined by comparing all of the 3D shapes, show low value, indicating the high-dimensional precision of the component. Consequently, it can be said that this work contributes to sectors where product differentiation is a significant necessity. Using single point incremental forming, a straightforward frame that grips the sheet during the operation is used to position the sheet for shaping on the table of a three-axis CNC machine. A simple tool is used to continuously move in line with an automated trajectory, which is developed using the CAD/CAM system, to build the product. The tangential motion of the tool in the current analysis depicts a spiral form that progressively falls in the Z-direction from the outside of the shape to the interior; in this way, the tool path completely generates the product profile [38] (Figure 14.2).

14.3.2   Identified Challenges

The order platform's definition of words calls for taking into account the professional distinctions between clinicians and technologists. Here, the emphasis should be on how the application allows users to use common medical phrases to describe the intended result, and in the best scenario, how the application might transform this supplied information into the pertinent technical information needed to produce the 3D-printed item.

The other challenge linked to the previous one is making sure that the information supplied on the sorting of medical equipment is fully in accordance with the regulations on medical devices. This is necessary in order to identify the proper risk class of the medical equipment.

The extent to which the user should be able to provide particular technical information regarding the 3D printing process itself, such as the layer height or part positioning on the build plate, is a crucial final point to make. Determining these parameters demands an in-depth technical understanding of the 3D printing process since they have a major influence on the mechanical qualities of the finished product [39].

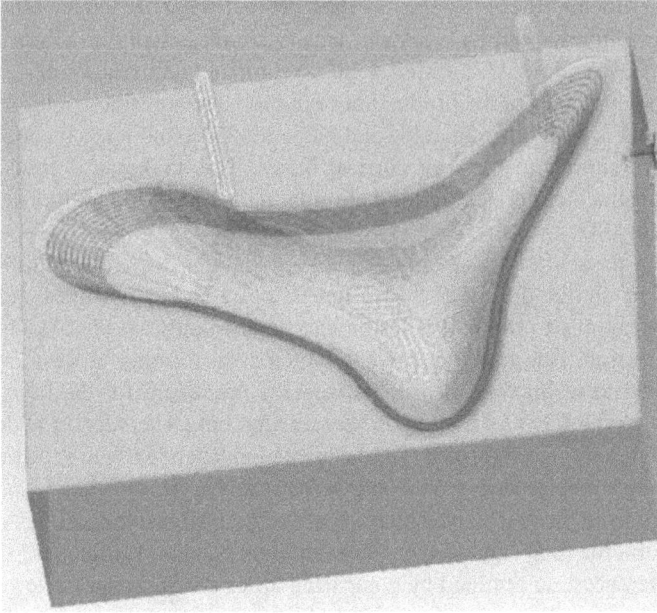

FIGURE 14.2   3D illustration of tool trajectory [38].

14.4   CONCLUSION

Continuous development and research in the field of medical implants widen the material selection that can be used for different applications inside the human body. They increase the life expectancy of implants as well as provide the probable solution to inflammatory discomfort and biotoxicity with tissues. This chapter acknowledges the different biocompatible materials and manufacturing processes developed in recent times for the production of medical implants. Titanium and its alloy are most common and widely used for bioimplants; however, they cause allergic reactions due to the separation of the metal ions in tissues. Ceramic implants are also used as medical implants due to their excellent inertness with body tissues. Generally, aluminum oxide (Al_2O_3) and/or zirconium dioxide (ZrO_2) are used as inserts for dental implants and prostheses. Polymer implants are receiving more attention in biomedical implants due to their lightness, high flexibility, and ease of manufacture. However, these are limited by poor interchangeability, strength, and stability. The chapter also discusses the manufacturing techniques used to produce medical implants. The manufacturing techniques for medical implants greatly depend on the specific application and implant position inside the body. Additive manufacturing is the most popular technique for the development of patient-oriented implants. The development of metal additive manufacturing opens the door for metallic implants such as titanium, cobalt, and tantalum. Other manufacturing techniques discussed are incremental forming (SPIF) and sintering.

REFERENCES

1. Patel, N., P. Gohil. A review on biomaterials: Scope, applications & human anatomy significance. *International Journal of Emerging Technology and Advanced Engineering* 2 (2012) 91–101. http://citeseerx. ist.psu.edu/viewdoc/download;jsessionid=1A91665506679B 581C5915B94776C0F2? doi=10.1.1.413.7368&rep=repl&type=pdf.

2. Geetha, M., A.K. Singh, R. Asokamani, A.K. Gogia. Ti based biomaterials, the ultimate choice for orthopaedic implants - A review. *Progress in Materials Science* 54(3) (2009), 397–425, https://doi.org/10.1016/j.pmatsci.2008.06.004.

3. Shabalovskaya, S., J. Anderegg, J. Van Humbeeck. Critical overview of Nitinol surfaces and their modifications for medical applications. *Acta Biomaterialia* 4(3) (2008), 447–467, https://doi.org/10.1016/j.actbio.2008.01.013.

4. Kirilova, I.A., M.A. Sadovoy, V.T. Podorozhnaya, S.P. Buyakova, S.N. Kulkov. Ceramic and osteoceramic implants: Upcoming trends. *Hir. Pozvonočnika (spine surgery)* (2013), 52–62, https://doi.org/10.14531/ss2013.4.52-62.

5. Andreiotelli, M., H.J. Wenz, R.-J. Kohal. Are ceramic implants a viable alternative to titanium implants? A systematic literature review. *Clinical Oral Implants Research* 20 (suppl 4) (2009), 32–47, https://doi.org/10.1111/j.1600-0501.2009.01785.x.

6. Petersen, R. Carbon fiber biocompatibility for implants. *Fibers* 4(1) (2016), 1, https://doi.org/10.3390/fib4010001.

7. Nazerali, R., J. Rogers, R. Canter, K.M. Hinchcliff, T.R. Stevenson. The use of polypropylene mesh in chest wall reconstruction; A novel approach. *Journal of Plastic, Reconstructive and Aesthetic Surgery* 68(2) (2015), 275–276, https://doi.org/10.1016/j.bjps.2014.09.036.

8. Mahabir, R.C., C.E. Butler. Stabilization of the chest wall: Autologous and alloplastic reconstructions. *Seminars in Plastic Surgery* 25(1) (2011), 34–42, https://doi.org/10.1055/s-0031-1275169.

9. Scholz, M.-S., J.P. Blanchfield, L.D. Bloom, B.H. Coburn, M. Elkington, J.D. Fuller, M.E. Gilbert, S.A. Muflahi, M.F. Pernice, S.I. Rae, J.A. Trevarthen, S.C. White, P.M. Weaver, I.P. Bond. Review. *Composites Science and Technology* 71(16) (2011), 1791–1803, https://doi.org/10.1016/j.compscitech.2011.08.017.

10. Li, C.S., C. Vannabouathong, S. Sprague, M. Bhandari. The use of carbon-fiber reinforced (CFR) PEEK material in orthopedic implants: A systematic review. *Clinical Medicine Insights: Arthritis and Musculoskeletal Disorders* 8 (2015), 33–45, https://doi.org/10.4137/CMAMD.S20354.

11. Basova et al. The use of noble metal coatings and nanoparticles for the modification of medical implant materials. *Materials and Design* 204 (2021), 109672.

12. Balažic, M., D. Recek, D. Kramar, M. Milfelner, J. Kopač. Development process and manufacturing of modern medical implants with LENS technology. *Journal of Achievements in Materials and Manufacturing Engineering* 32(1) (2009), 46–52.

13. Gebhardt, A., F.M. Schmidt, J.S. Hötter, W. Sokalla, P. Sokalla. Additive manufacturing by selective laser melting the realizer desktop machine and its application for the dental industry. *Physics Procedia* 5 (2010), 543–549.

14. Novakov, T., M.J. Jackson, G.M. Robinson, W. Ahmed, D.A. Phoenix. Laser sintering of metallic medical materials—A review. *International Journal of Advanced Manufacturing Technology* 93(5–8) (2017), 2723–2752.

15. Zhang, E., F. Li, H. Wang, J. Liu, C. Wang, M. Li, K. Yang. A new antibacterial titanium–copper sintered alloy: Preparation and antibacterial property. *Materials Science and Engineering: Part C* 33(7) (2013), 4280–4287.

16. Wu, H., X. Zhang, X. He, M. Li, X. Huang, R. Hang, B. Tang. Wear and corrosion resistance of anti-bacterial Ti–Cu–N coatings on titanium implants. *Applied Surface Science* 317 (2014), 614–621.

17. Muhamedov, M., D. Kulbakin, V. Gunther, E. Choynzonov, T. Chekalkin, V.Hodorenko. Sparing surgery with the use of TiNi-based endografts in larynxcancer patients. *Journal of Surgical Oncology* 111(2) (2015), 231–236, https://doi.org/10.1002/jso.23779.

18. Basova, Tamara V., et al. The use of noble metal coatings and nanoparticles for the modification of medical implant materials. *Materials and Design* 204 (2021), 109672.

19. Best, S.M., et al. Bioceramics: Past, present and for the future. *Journal of the European Ceramic Society* 28(7) (2008), 1319–1327.

20. Clarke, Ian C. Role of ceramic implants. Design and clinical success with total hip prosthetic ceramic-to-ceramic bearings. *Clinical Orthopaedics and Related Research (1976–2007)* 282 (1992), 19–30.

21. Vallet-Regí, Maria. Ceramics for medical applications. *Journal of the Chemical Society, Dalton Transactions* 2(2) (2001), 97–108.

22. Barinov, Sergei M. Calcium phosphate-based ceramic and composite materials for medicine. *Russian Chemical Reviews* 79(1) (2010), 13.

23. Chen, Qizhi, Chenghao Zhu, George A. Thouas. Progress and challenges in biomaterials used for bone tissue engineering: Bioactive glasses and elastomeric composites. *Progress in Biomaterials* 1(1) (2012), 1–22.

24. Bayazit, Vahdettin, Murat Bayazit, ElifBayazit. Evaluation of bioceramic materials in biology and medicine. *Digest Journal of Nanomaterials and Biostructures* 7(2) (2010), 267–278.

25. Kirilova, I.A., M.A. Sadovoi, V.T. Podorozhnaya. Ceramic and bone-ceramic implants: Promising directions. *KhirurgiyaPozvonochnnika* 4 (2013), 52–62.

26. De Aza, Piedad N., et al. Bioeutectic® ceramics for biomedical application obtained by Laser Floating Zone method. In vivo evaluation. *Materials* 7(4) (2014), 2395–2410.

27. Ayawanna, Jiratchaya, Namthip Kingnoi, Nattapol Laorodphan. A feasibility study of egg shell-derived porous glass–ceramic orbital implants. *Materials Letters* 241 (2019), 39–42.

28. Osés, Javier, et al. Antibacterial functionalization of PVD coatings on ceramics. *Coatings* 8(5) (2018), 197.

29. Ohtsuki, Chikara, Masanobu Kamitakahara, Toshiki Miyazaki. Bioactive ceramic-based materials with designed reactivity for bone tissue regeneration. *Journal of the Royal Society. Interface* 6 (suppl)_3 (2009), S349–S360.

30. Matsko, A., R. França. Design, manufacturing and clinical outcomes for additively manufactured titanium dental implants: A systematic review. *Dentistry Review* (2022), 100041.

31. Rifai, A., S. Houshyar, K. Fox. Progress towards 3D-printing diamond for medical implants: A review. *Annals of 3D Printed Medicine* 1 (2021), 100002.

32. Salmi, M. Additive manufacturing processes in medical applications. *Materials* 14(1) (2021), 191.

33. Meena, V.K., P. Kumar, P. Kalra, R.K. Sinha. Additive manufacturing for metallic spinal implants: A systematic review. *Annals of 3D Printed Medicine* 3 (2021), 100021.

34. Aufa, A.N., M.Z. Hassan, Z. Ismail. Recent advances in Ti-6Al-4V additively manufactured by selective laser melting for biomedical implants: Prospect development. *Journal of Alloys and Compounds* 896 (2022), 163072.

35. Jiao, L., Z.Y. Chua, S.K. Moon, J. Song, G. Bi, H. Zheng. Femtosecond laser produced hydrophobic hierarchical structures on additive manufacturing parts. *Nanomaterials* 8(8) (2018), https://doi.org/10.3390/nano8080601.

36. Li, C., D. Pisignano, Y. Zhao, J. Xue. Advances in medical applications of additive manufacturing. *Engineering* 6(11) (2020), 1222–1231.
37. Palumbo, G., A. Cusanno, M.G. Romeu, I. Bagudanch, N.C. Negrini, T. Villa, S. Farè. Single Point Incremental Forming and electrospinning to produce biodegradable magnesium (AZ31) biomedical prostheses coated with porous PCL. *Materials Today: Proceedings* 7 (2019), 394–401.
38. Ambrogio, G., L. De Napoli, L. Filice, F. Gagliardi, M. Muzzupappa. Application of Incremental Forming process for high customised medical product manufacturing. *Journal of Materials Processing Technology* 162 (2005), 156–162.
39. Url, P., D. Stampfl, M. Tödtling, W. Vorraber. Challenges of an additive manufacturing service platform for medical applications. *Procedia CIRP* 112 (2022), 400–405.

Conclusion

Biomedical implants play a vital role in improving the quality of life for patients with a wide range of medical conditions. As technology continues to advance, we can expect further improvements in implant materials, design, and reliability, leading to enhanced patient outcomes and reduced healthcare costs.

However, it is crucial to recognize that the development and implementation of biomedical implants must be accompanied by careful consideration of regulatory and ethical aspects. Balancing patient safety, informed consent and equitable access to these life-changing technologies is essential for the future of healthcare. By offering a balanced perspective on both the benefits and challenges associated with biomedical implants, this book equips readers with a solid understanding of this rapidly evolving field. Whether you are a healthcare professional, researcher, or student, this book serves as a valuable resource for gaining insights into the current state-of-the-art and future prospects of biomedical implants. We hope that it sparks further innovation, fosters interdisciplinary collaborations, and contributes to the ongoing advancement of patient care and quality of life through these remarkable technologies. It provides a foundation of knowledge and insights that can guide future advancements and contribute to the development of safe, effective, and patient-centered implantable medical devices.

DOI: 10.1201/9781003375098-15

Index

Page number followed by *f* indicate figure and **bold** indicate table respectively

For Product Safety Concerns and Information please contact our EU
representative GPSR@taylorandfrancis.com
Taylor & Francis Verlag GmbH, Kaufingerstraße 24, 80331 München, Germany